The Cellular Telephone Installation Handbook

Michael Losee

QUANTUM PUBLISHING, INC.

Acknowledgments

We want to offer special thanks and acknowledgment to Gina M. Passaretti, and to Marc DeLong and Tracy Dybowski for reading the manuscript and offering suggestions and additions. And to Melissa Macaluso for her help with the photographs.

We also wish to thank the following companies for providing the photographs as listed: Nokia-Mobira, pp. 7, 40, 58, 59; Panasonic Industrial Co., p. 7; NEC America, p. 7; Plexsys Corp., pp. 21, 24; Marconi Instruments, p. 29; Curtis Electro Devices, p. 39; Uniden Corp., p. 56; General Electric Co., pp. 56, 58; Audiovox Corp., p. 56; Motorola, Inc., pp. 58, 59; Mitsubishi International, p. 59; Morrison & Dempsey Communications, p. 75; Spectrum Cellular, pp. 76, 163; The Antenna Specialists Co., pp. 88, 93, 111; ORA Electronics, pp. 91, 103-8, 112, 119, 133, 138-39; PanaVise Products, Inc., p. 137; and Solarex, p. 159.

Published by Quantum Publishing, Inc.
Mendocino, California 95460

© 1988 by Quantum Publishing, Inc.
First edition published September 1988.

All rights reserved under international and Pan-American copyright conventions. No portion of this book may be reproduced by any means whatsoever, except for brief quotations in reviews, without written permission from the publisher.

Printed in the United States of America.

Library of Congress Cataloging-in-Publication Data

Losee, Michael, 1959–
   The cellular telephone installation handbook / by Michael Losee.
     p.   cm.
   Includes index.
   ISBN 0-930633-05-9 : $49.95
   1. Cellular radio—Handbooks, manuals, etc.   I. Title.
TK6570.M6L67 1988
621.3845—dc19

                                                   88-18325
                                                      CIP

To Mark and Jan Mancini, who taught me the business.

Contents

Foreword xi
Preface xiii

CHAPTER 1 A HISTORY OF CELLULAR TELEPHONES 1

CHAPTER 2 THEORY AND OPERATION OF A CELLULAR SYSTEM 5
System Overview 5
Cellular Geographic Service Areas 8
Call Processing 9
Handoff 11
Directed Retry 12
Channel Seizure 14
Mobile Terminal 14
Dropped Calls 15
Multipath Interference 16
Signaling 17
Electronic Serial Number Verification 18
Roaming 18
Landline Interfaces 19

MTSO Description 19
> *Functions • Hardware • System Software*

Cell-Site System Description 23

CHAPTER 3   TEST EQUIPMENT AND TOOLS 27

Test Equipment 28
> *Cellular Test Center • Watt Meter • NAM Programming • NAMs (Number Assignment Modules) • Power Supply • Test Light • Multimeter*

Tools 42

Facilities 47

Computer Equipment 49

CHAPTER 4   CELLULAR TELEPHONE EQUIPMENT 55

Cellular Telephones 55
> *Mobile-Only Cellular Telephones • Transportable Cellular Telephones • Hand-Held Portable Cellular Telephones • Cellular Telephone Features*

Add-On Accessories 74

Carrier Services 77

CHAPTER 5   ANTENNA THEORY AND SELECTION 79

Antenna Theory 79
> *Wavelength • Antenna System • Antennas • Antenna Radiation Patterns*

Antenna Selection 88
> *Glass-Mount Antennas • Elevated-Feed Antennas • Roof-Top Antennas*

The Losee-Shosteck Effect 96

CHAPTER 6   VEHICLE INSTALLATIONS 99

Prechecking the Vehicle 99

Planning the Installation 102
> *Selecting the Control Head Location • Selecting the Transceiver Location • Selecting the Antenna Location*

Beginning the Installation 114
Security Radios • On-Board Computers • A Word of Caution • Passive Restraint Systems
Disconnecting the Battery 116
Connecting the Power Cable 116
Ground Wire • Running the Power and Data Cables • Ignition Sense Wire • Constant Power Wire
Installing the Antenna 124
Glass-Mount Antennas • Elevated-Feed Antennas • Roof-Top Antennas
Testing and Programming the Equipment 130
Auto Test • Manual Handoff Test
Mounting the Transceiver 132
Mounting the Control Head 135
Installing the Hands-Free Microphone 138
Selecting a Location • Running the Cable • Calibrating the Hands-Free Microphone
Connecting the Data Cable 140
Making a Test Call 140
Checking Out the Customer 141
Review of Installation Procedures 141

CHAPTER 7 **MARINE INSTALLATIONS 143**
Planning the Installation 144
Marine Antennas 144
Theory of Operation • Applications
Selecting the Antenna Location 145
Sailboats • Powerboats
Installing the Antenna 147
Sailboats • Powerboats
Mounting the Control Head 148
Mounting the Transceiver 149

Completing the Installation   150
Add-On Accessories   150

CHAPTER 8   RURAL INSTALLATIONS   153

Cellular Telephone   154
Antenna   155
Solar Panels   158
Voltage Regulator   159
Batteries   159
RJ11 Interface   160
Computer Interfaces   162
Installing the Fixed Installation   162
   Antenna • Running the Cable • Solar Panels • Utility Closet Equipment

CHAPTER 9   TROUBLESHOOTING AND REPAIRING INSTALLATIONS   169

Cellular Telephone Subsystems   170
The Phone Turns On, But Will Not Place or Receive a Call   171
The Phone Usually Places Calls, But Not Always   175
The Phone Will Not Turn On   176
The Phone Drops Too Many Calls   180
The Phone Needs to Be Unlocked   189
The Phone Turns On and Off by Itself   190
Failure Due to Age and Use   191
The Hands-Free Unit Sounds Terrible   192
Repairing a Hole in the Gas Tank   193
Repairing a Hole in the Fuel Line   195
Summary of Troubleshooting and Repair Problems   196

APPENDIX A   GLOSSARY   201

APPENDIX B   DIRECTORY OF MANUFACTURERS   209

Index   229

Foreword

The Cellular Telephone Installation Handbook marks a transition in the cellular industry from its early stage of wild and erratic growth to a maturing period of planned and sustained expansion.

This is as it should be. Cellular, as with any new industry, has had to experience the pains of birth and the mistakes of maturation. While the latter are not over, the industry today has more of a coherence and sense of direction that was often lacking in the past. There is an increasing recognition that for the industry to prosper, the manufacturer, the carrier, and the dealer alike must cooperate among themselves to assure maximum service and satisfaction for the end-user customer.

The Cellular Telephone Installation Handbook marks this recognition. It will be distributed primarily by manufacturers and carriers. This will attest to their acknowledgment of the central role of the telephone and antenna installation in ensuring the provision of satisfactory cellular service to the end user.

Poor telephone and antenna installations, particularly the latter, are responsible for anywhere from 70 to 90 percent of customer complaints concerning poor or inadequate service. The costs of such poor installations to dealers, manufacturers, and system operators have been enormous. Dealers have spent hours correcting improperly done initial installations. Manufacturers have spent vast sums trading out transceivers that, upon inspection, have been found to be in good working order. System operators have suffered low use, slow pay, churn, and switch overloads due to poorly installed telephones that caused poor service, even with the best of system designs. For all of these reasons, greater attention to the nuts and bolts of proper telephone and antenna installation has been long overdue.

The Cellular Telephone Installation Handbook provides this attention. Mike Losee is uniquely qualified to write it. He brings to this task three essential perspectives. First, he has the theoretical perspective of a trained radio engineer who can place the issues of telephone installation into a broader perspective. Second, he has the hands-on experience of an agent/dealer who has dealt with the problems of "real world" installations on a daily basis. Third, he can convey this experience and knowledge in a clear and easily understood style.

The Cellular Telephone Installation Handbook is a landmark book. On the one hand, it marks a symbolic maturation of the cellular industry. On the other hand, it is a practical tool that will help to see the industry through that maturation period.

HERSCHEL SHOSTECK

Preface

In the early 1980s, when the first cellular system went into service, several questions plagued this emerging industry. One of the more difficult questions was, "Who is going to install all of these cellular telephones?" Most of the cellular service carriers decided that selling and installing these new phones was more than they could handle, considering that they had the huge task of engineering and building the cellular systems. Most of the carriers decided to appoint agents to sell and install this new product.

The focus of this book is to teach the skill of installing cellular telephones. While the installation of a cellular telephone is not a difficult process, it is one that must be learned. During the early days of cellular, it seems a lot of phones were destroyed and a lot of gas tanks were drilled by unskilled people claiming to be installation "experts."

The Cellular Telephone Installation Handbook is a step-by-step guide that teaches the latest and most effective installation, troubleshooting, and repair techniques. Antenna theory, selection, placement, and installation are explained, along with complete sections on cellular system theory; test equipment and tools; telephone selection, placement, and installation; marine and rural (fixed) installations; and procedures for dealing with customers from check-in through check-out and followup.

All the information is presented in an easy-to-understand manner. New technical terms are either explained in the text as they come up or included in the glossary. More than 100 diagrams, charts, and photographs provide an informative visual presentation of the material. Additionally, for your convenience, we have included a directory of equipment manufacturers, distributors, and service providers.

Loaded with tips and tricks of the trade that would otherwise take years to learn, all the information necessary to produce perfect installations (which, in turn, create satisfied customers) is supplied to you in this book. Remember, satisfied customers create the best advertising a company can get: positive customer referrals.

Good installers are hard to find in today's market. This is a new industry, and the need for qualified people is exceptional. Using *The Cellular Telephone Installation Handbook*, readers will become qualified to start their own service center, be hired as an installation technician, and/or improve their skills as a marketer of cellular products and services.

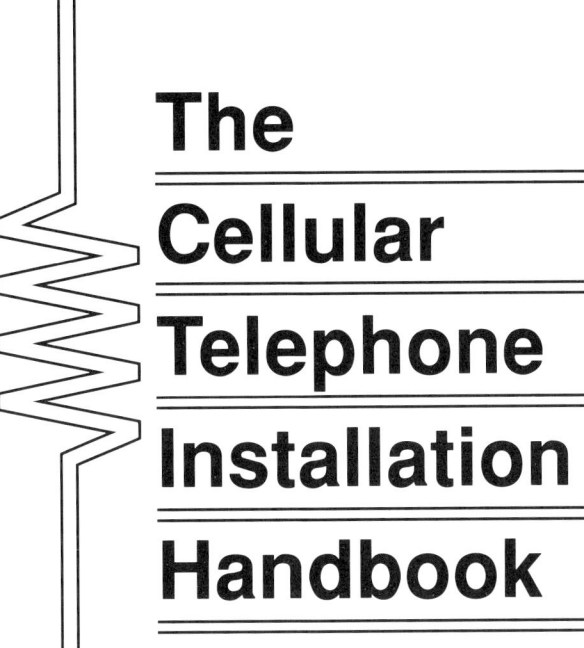

The Cellular Telephone Installation Handbook

relatively low frequencies were used. These factors allowed the system operators to cover the largest areas for the least amount of money.

The system operators were entities (or companies) that were licensed to provide mobile telephone service, much like the cellular carriers of today. In most major cities, service was provided by the local Bell Telephone company and a few private companies, sometimes called radio common carriers (or RCCs). Customers had the choice of selecting the carrier that provided the service they needed. However, the service was expensive, and the quality of the connection was poor.

The hardware associated with this technology was massive by today's standard. The mobile units could weigh 20 or 30 pounds and consume 30 or so amperes while in use. If the stereo was on, the "spike" on the vehicle's DC power supply would almost destroy the speaker. I remember an old IMTS unit I had installed in a small sports car. It was easy to tell when the mobile telephone was about to ring because the vehicle's headlights would dim under the extensive current drain.

The mobile units operated on the 150-MHz (and later on the 450-MHz) band and radiated about 50 watts. A single antenna provided the coverage for each city. This restricted a system operator to using a particular channel only once per city. The mobile units were very expensive and the audio quality was not very good, but they did provide mobile telephone communications. The problem was the shortage of conversation channels. The Federal Communications Commission (FCC) had allocated only a few dozen channels for even the largest cities. They had not anticipated the great demand for mobile telephone service.

The system grew like roads in an old city—it handled traffic, but not very well. This was fine for small cities where perhaps only a few hundred people were interested in the service (particularly at the high rates being charged), but in large metropolitan areas, it was woefully inadequate. With few channels available, only a few conversations were possible at one time. I remember driving through New York City in the early 1980s and attempting to place a call on an IMTS telephone that was equipped with "illegal" channel-grabbing circuitry. This gave us an unfair advantage over other people trying to place calls, and we still had to wait half an hour before we could place a call.

In the late 1950s, the people at Bell Laboratories began to design a system to replace IMTS as the standard for mobile telephony. This system needed to be able to handle thousands of calls at one time. The billing needed to be automatic, and the audio quality of the conversation required considerable improvement. The ability to place and receive calls was also imperative. The concept required high-speed signaling and sophisticated computer logic that did not exist at the time. It was almost twenty years before anyone was sure if the technology was feasible. The result was the Advanced Mobile Phone Service, or AMPS. It was to provide all these features after languishing with the FCC for almost fifteen years.

When the FCC gave AMPS (which by now was called *cellular*) approval, 666 channels were immediately allocated for the new service. Additional spectrum to bring the total number of channels up to 832 was held in reserve. Now, system operators in large cities had an almost (and almost is the operative word) unlimited ability to sell mobile telephone service.

CHAPTER 2

Theory and Operation of a Cellular System

Before you learn how to set up a service shop or how cellular telephone installations are accomplished, it's a good idea to have a basic understanding of cellular theory and operation.

SYSTEM OVERVIEW

A cellular system consists of several building blocks assembled to form a working communications network. The three basic building blocks are

1. The Mobile Telephone Switching Office (MTSO)
2. A cell site (A system will support multiple cell sites.)
3. A cellular terminal

Figure 2.1 illustrates a typical cellular system. A cellular telephone system is a fully automatic, stored-program device. (A stored-program device is one that is controlled by a computer program.) It is designed to provide cellular telephone service to communities using fixed, portable, and mobile terminals. A properly designed cellular system allows communications between mobile/portable units and POTS (plain old telephone service) devices, other cellular units, and other electronic devices (such as computers, data networks, voice mail and messaging systems, etc.) that are capable of interfacing with the public telephone switching network (PTSN). Figure 2.2 shows the three types of cellular telephones: mobile-only, transportable, and hand-held portable.

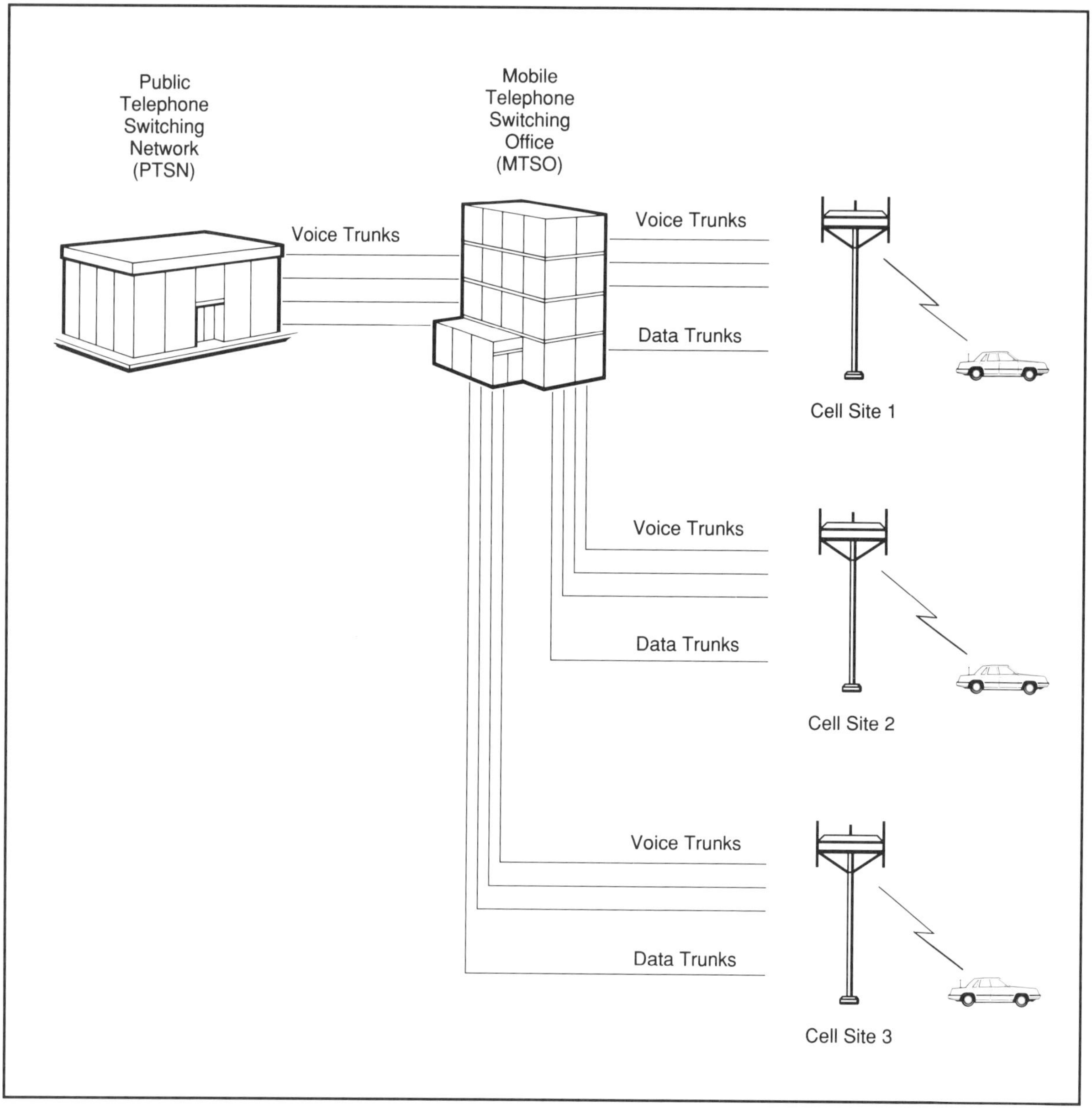

6 THEORY AND OPERATION OF A CELLULAR SYSTEM

FIGURE 2.1 CONFIGURATION OF A TYPICAL CELLULAR SYSTEM
Cellular terminals are interfaced with the landline telephone network (PTSN) through the MTSO.

The cell sites communicate with the cellular telephone equipment via relatively low-powered 800-MHz transmitters. The power output of a cell site cannot exceed 100 watts ERP (effective radiated power). The number and location of cell sites are determined by specific traffic and geographic needs.

THEORY AND OPERATION OF A CELLULAR SYSTEM 7

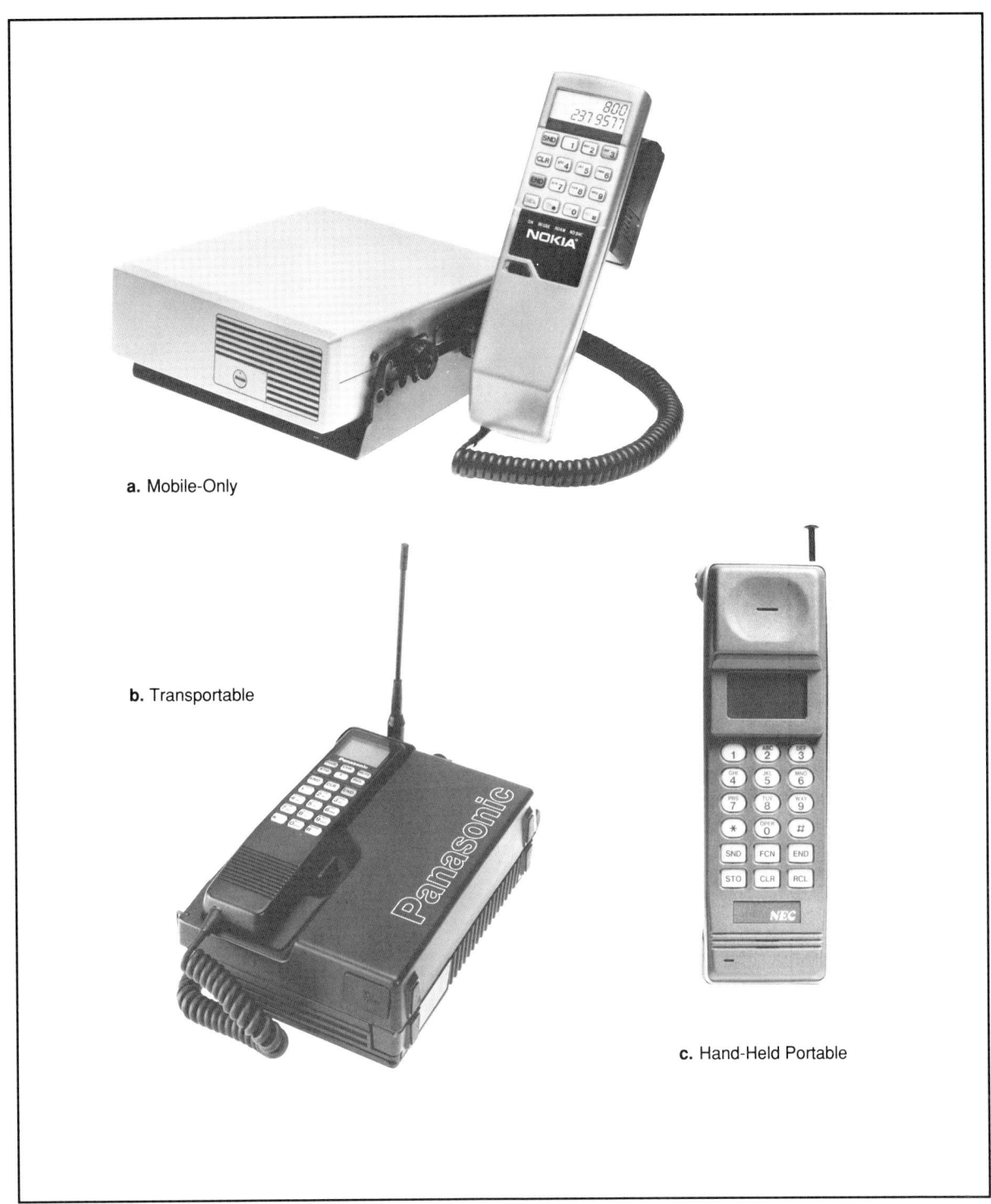

a. Mobile-Only

b. Transportable

c. Hand-Held Portable

FIGURE 2.2 TYPES OF CELLULAR TELEPHONES

8 THEORY AND OPERATION OF A CELLULAR SYSTEM

CELLULAR GEOGRAPHIC SERVICE AREAS

The intent of a cellular system is to provide continuous geographical coverage within a Cellular Geographic Service Area, or CGSA. Figure 2.3 shows the New York City wireline carrier's CGSA. When the FCC approved the framework for cellular telephony, it divided service areas into CGSAs. The purpose was to define "communities of interest" surrounding major metropolitan areas and to ensure continuous communications throughout each of these areas. An example of this is the New York City CGSA. This service area was designed to include northern New Jersey, southern New York state, and parts of Long Island. The logic was that these areas are related to each other economically and geographically. Many people who live in northern New Jersey and southern New York state commute to or have business

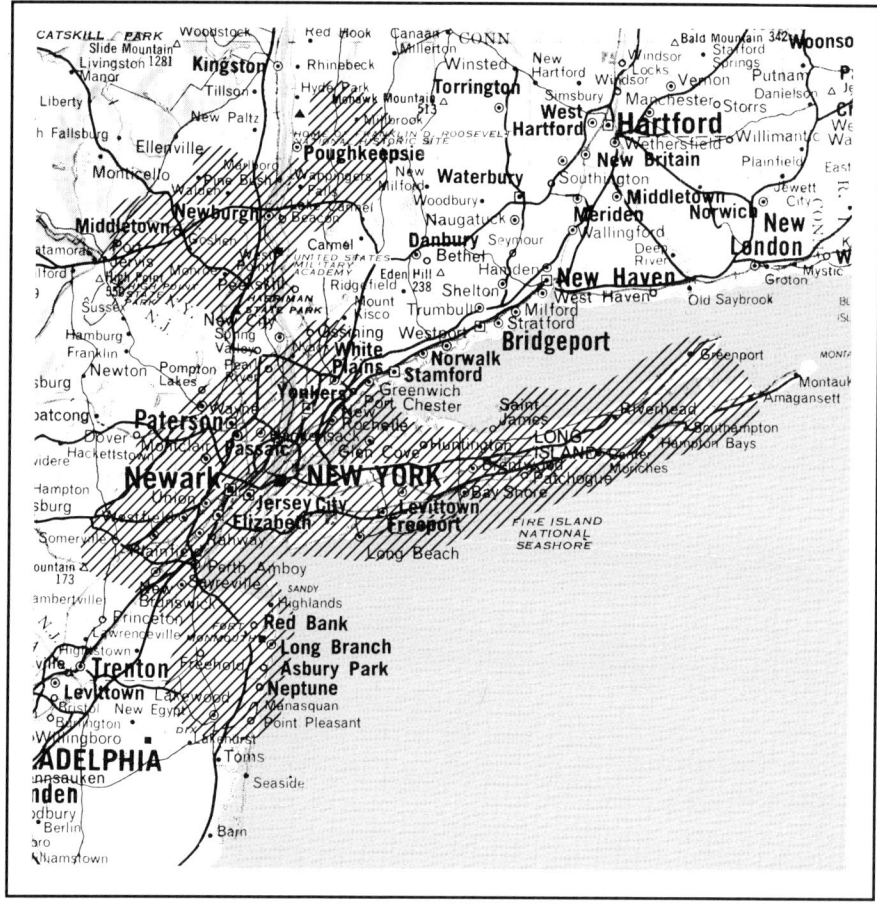

FIGURE 2.3 NEW YORK CITY WIRELINE CARRIER'S CGSA

relationships in New York City. It was logical to allow continuous service in these areas.

These "communities of interest" also created some problems for the service providers. The New York City CGSA covers five Number Planning Areas (NPAs), or area codes, and two states (southern New York and northern New Jersey). While the division of areas into CGSAs created many regulatory and technical problems during the initial system setup, the cellular subscriber was ultimately better served by it.

Communications between mobile terminals and cell sites is accomplished by using a set of preassigned radio transceivers within the 800-MHz frequency band. These radio transceivers provide a radio frequency link between the mobile terminals and the cell sites. A portion of these radio transceivers are reserved for paging and access. The remaining radio channels are used as voice channels. The number of voice, paging, and access channels available depend upon the local traffic demands and the grade of service the carrier wishes to provide. Most carriers offer P02 service. This means that during peak usage periods only 2 percent of all call attempts will be blocked on the first attempt.

P02 is considered to be an industry standard for an acceptable grade of service. As systems become more crowded with additional subscribers, the service providers will be forced to reduce their grade of service since they have finite amounts of spectrum available to them. Because only one conversation can take place on a channel at any particular time, some blocking (inability to find a free channel) may occur during periods of heavy usage.

CALL PROCESSING

The mobile terminal is programmed to monitor the control channels and to "lock" onto the control channel with the strongest signal. Only a small number of the total 832 channels allocated for cellular service are designated as control channels. Since these channels are used only to "set up" a cellular call, a mobile terminal remains "locked" onto a control channel for only a short period of time. This allows a large number of mobile terminals to be served by a relatively small number of control channels. Mobile terminals are designed to repeat this scanning process periodically, since the relative signal strength of various control channels will change as the mobile terminal changes its physical position. Figure 2.4a shows a cellular subscriber (terminal) that has locked

10 THEORY AND OPERATION OF A CELLULAR SYSTEM

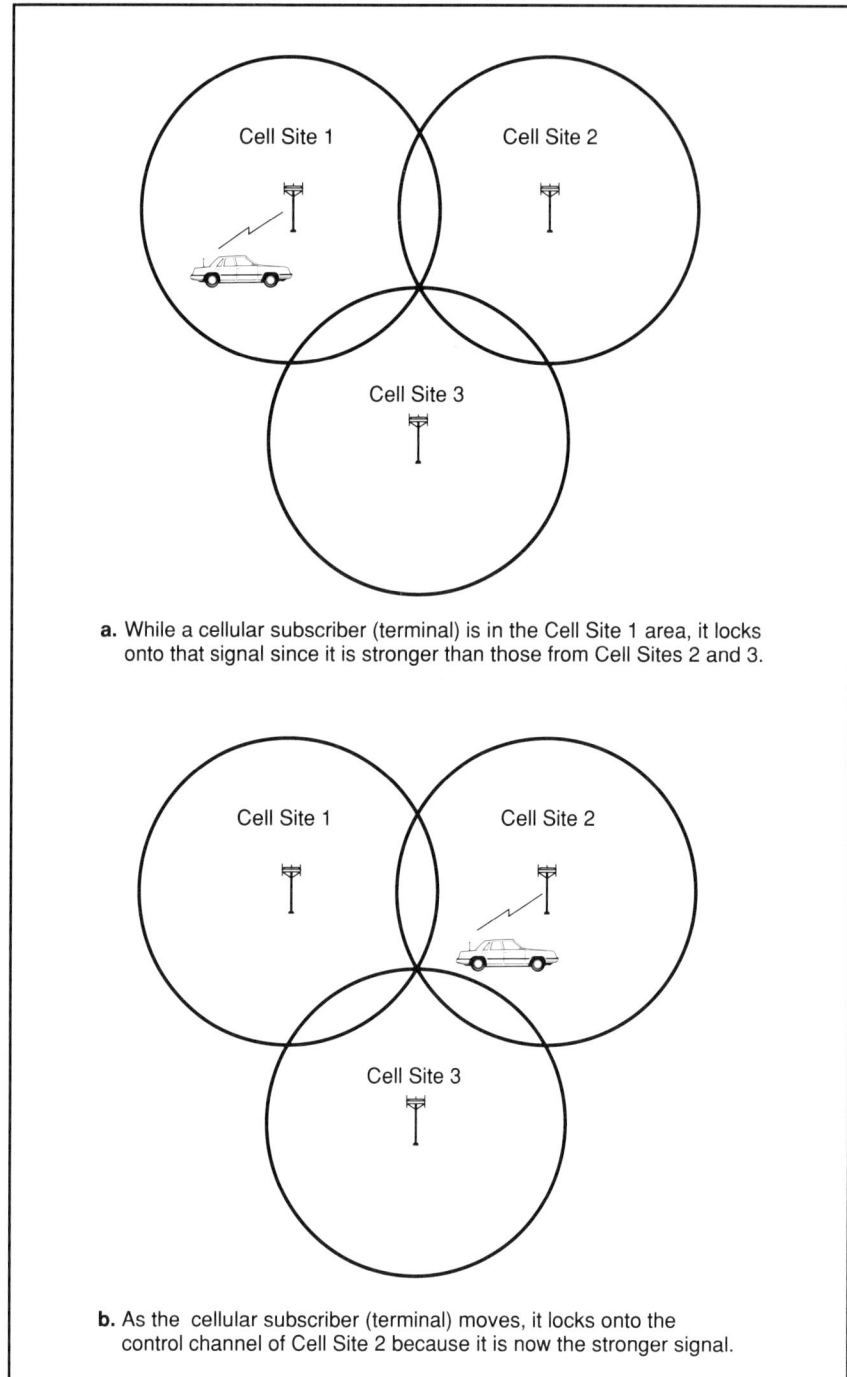

a. While a cellular subscriber (terminal) is in the Cell Site 1 area, it locks onto that signal since it is stronger than those from Cell Sites 2 and 3.

b. As the cellular subscriber (terminal) moves, it locks onto the control channel of Cell Site 2 because it is now the stronger signal.

FIGURE 2.4 PAGING AND ACCESS

onto its nearest cell site. This ensures that the mobile terminal will communicate with the cell site whose signal is strongest. Strong, noise-free channels are essential for clear, error-free communication. The control channels are used to exchange data between the cell site and the mobile terminal. This data includes but is not limited to

1. The called-party telephone number
2. The calling-party telephone number
3. The calling-party electronic serial number

Voice communications do not take place on control channels. After the mobile terminal "locks" onto or seizes a control channel, it monitors the data stream for relevant information. When a mobile terminal receives a page from a cell site, it rescans the control channels before sending its response to ensure that it will be communicating with the "best" cell site. In figure 2.4b, the mobile terminal has rescanned and locked onto the control channel of cell site 2. After responding to a page from the cell site, the mobile terminal is then directed to change frequency to a voice channel.

HANDOFF

Once in communication with a cell site, the mobile terminal has the capability of moving anywhere within the CGSA where radio frequency (RF) coverage is sufficient. When a system user who is engaged in a conversation crosses a cell-site boundary, the MTSO is responsible for maintaining the conversation by initiating a *handoff*. The process of handing off a cellular telephone conversation is what makes this system architecture unique.

In a properly designed cellular system, a mobile terminal usually has a choice of a number of cell sites with which it can communicate at any given time. This is done to reduce the chance of dropped calls due to local interference and to improve the traffic-handling capabilities of the system. In figure 2.5, the terminal may be served by either cell site 1, 2, or 3. Cell site 2 has been added to improve system traffic-handling capacity. The MTSO will determine which cell site will provide service for the terminal. In such a properly designed system, two mobile terminals physically located next to each other may be operating on different cell sites.

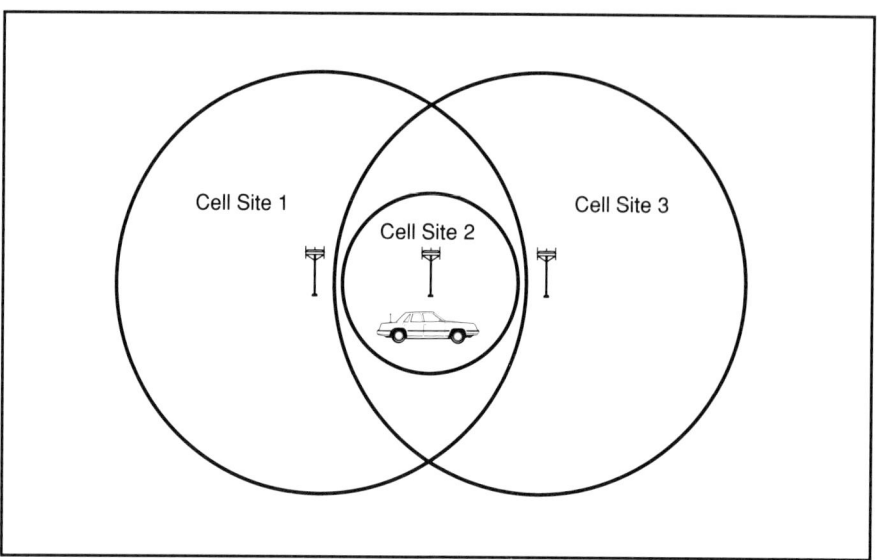

FIGURE 2.5 COVERAGE BY MULTIPLE CELL SITES
The MTSO can choose from among three cell sites to serve the terminal. Cell site 2 adds traffic-handling capacity to the system.

The MTSO is responsible for monitoring the mobile terminal's relative signal strength. Since it would not be practical to continually monitor each mobile unit, the MTSO polls or checks at regular intervals. This interval is normally between 5 and 10 seconds, depending on traffic conditions. When the signal strength (or SAT signal-to-noise ratio) falls below a certain level, the MTSO begins to search for an appropriate cell to handoff the call to.

Each cell site maintains a list of appropriate cells to handoff a call to. By measuring the mobile terminal's signal level at each of these cell sites, the MTSO can determine the best cell to handoff the call to (see figure 2.6).

DIRECTED RETRY

If the "best" cell site cannot accept the call because of traffic conditions or technical problems, the MTSO can use a technique called *directed retry*. This is a very important concept in larger systems where caller traffic can become so heavy that all of the channels in some cell sites are in use. Without this technique, the call would simply be dropped because there would be no place for it to be handed off to. When a cell site has all of its voice channels

THEORY AND OPERATION OF A CELLULAR SYSTEM   13

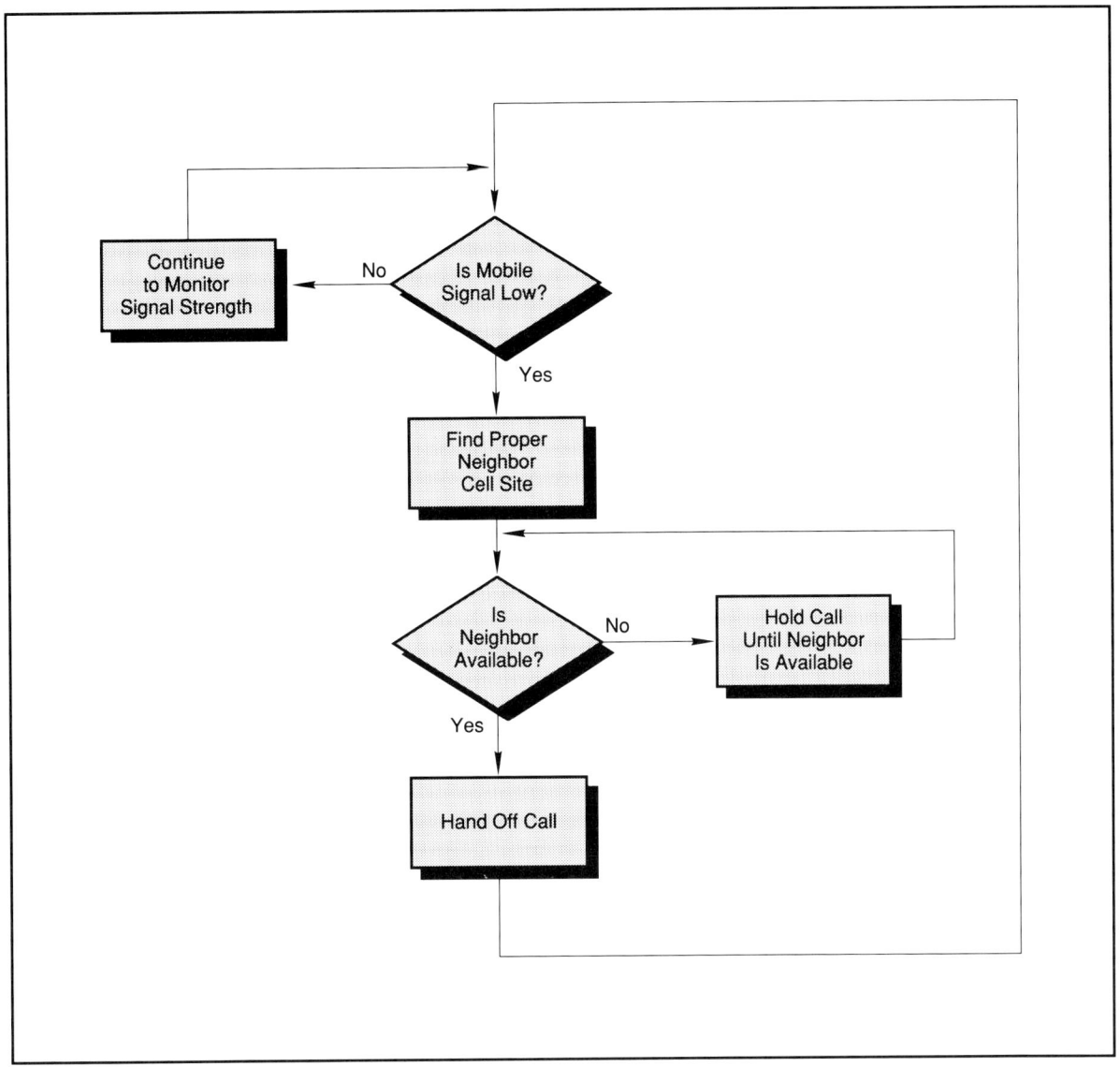

FIGURE 2.6 HANDOFF DECISION FLOWCHART

in use, it is not capable of accepting any additional conversations: it is not possible to hand off calls to that cell site.

The MTSO maintains a list of nearby cells, called a *neighbor list*. When a cell is full, the MTSO can hand off a call to a neighboring cell, thus eliminating a dropped call. The conversation is then maintained, even if the voice quality is a bit lower than normal (since the call is being handed to a cell that is not the "best" choice). By moving traffic in this way, the MTSO is able to handle peak service demand without adding additional hardware. In systems where additional channels are no longer available, directed retry can significantly improve the grade of service and reduce the number of dropped calls. One should remember that while the chance of dropping a call during a directed retry is higher than during a normal handoff, a net improvement in service does result from this technique.

CHANNEL SEIZURE

Channel seizure is the process of being assigned a voice channel by the MTSO. Cellular channel seizure functions in a manner similar to landline seizure. A POTS telephone has a dedicated line that is scanned by the central office for an off-hook condition. A properly designed landline system will not allow multiple seizures to occur on a single line. A cellular MTSO must ensure that multiple seizures do not occur on the same channel.

One of the techniques employed to reduce "channel collisions," or multiple seizures, is to assign the eleventh bit of the paging/access stream as the busy/idle bit. When a paging channel is in use, this bit is set to busy, and other mobile terminals will not attempt to seize this channel. The busy/idle bit is returned to the idle state after the channel is free. A waiting mobile terminal will wait for a pseudorandom (almost random) interval before attempting to access the channel again. This pseudorandom interval helps to prevent a number of mobile terminals from attempting to seize the paging channel at the same time during a period of heavy telephone traffic.

MOBILE TERMINAL

A mobile terminal attempting to seize a paging channel must identify which site it is trying to access. It does this by broadcasting a message containing the information. This is very important in mature cellular systems with small cells. Mobile terminals transmit at full power (3 watts at the antenna port or 6 watts ERP) during the paging portion of a conversation. By identifying the cell site with which the mobile terminal desires to communicate, co-channel seizures on adjacent cell sites are significantly reduced. A cell site is programmed to ignore a channel seizure intended for another cell site.

During the seizure process, the mobile terminal continues to monitor the condition of the busy/idle bit. If this condition becomes busy either too soon or too late, the mobile terminal is programmed to discontinue the call attempt and assumes the cell site is replying to another mobile terminal. In the event of an aborted call attempt, the mobile terminal will attempt to place the call several times. In order to prevent a paging-channel overload condition during heavy traffic periods (*remember:* there are only a few paging channels for thousands of mobile terminals), the mobile terminal will halt seizure attempts after a few times. The cellular subscriber is free to attempt to place a call after a short (a few seconds) timeout period.

DROPPED CALLS

It is during the handoff period that the mobile terminal is most vulnerable to a "dropped call." A *dropped call* is a cellular telephone call that is terminated against the wishes of the mobile or land party. A few of the causes are listed here.

1. Poor antenna placement or installation is the most frequent cause of a dropped cellular telephone call. A distorted RF pattern (which will be explained in Chapter 5, "Antenna Theory and Selection") or poor signal output cause data errors in communication between the mobile terminal and the MTSO. Customers complaining of a lot of dropped calls should have their antennas checked before anything else is done. (See the installation section for detailed information on correct antenna installation and placement.)

2. "Dead spots," or areas where the RF coverage from the cell site is inadequate, are another major cause. When a system is designed and implemented, there are inevitably areas where the RF coverage is poor. Uneven terrain, tall buildings, tunnels, bridges, interference, and, rarely, poor design can contribute to this problem. Dead spots most often occur in young systems and are corrected by the carrier as priorities permit.

3. Co-channel interference can cause dropped calls. This happens when a mobile terminal is able to detect the same channel from two different cell sites. The mobile terminal gets "confused" and usually resolves the conflict by dropping the call.

rates. Incomplete calls should be fewer than one in 10,000, and mispages or incorrectly identified users should be fewer than one in 10 million. To fight errors and meet these system requirements, all control messages are encoded using a technique called *BCH (Bose-Chaudhuri-Hocquenghem) coding*. Messages are also repeated several times. The receiver contains a ROM (read-only memory) look-up table; it can correct one error and detect two in each frame. This significantly reduces the error rate. The carrier is digitally modulated (turned into radio waves) using a method called *FSK (frequency shift keying)* with discriminator detection. Bits are encoded using a Manchester (a type of digital signaling) format.

ELECTRONIC SERIAL NUMBER VERIFICATION

One of the most important functions performed by the MTSO is electronic serial number (ESN) verification. Each manufacturer is assigned a three-digit code, which becomes the first three digits of an eleven-digit electronic serial number. Every cellular telephone is assigned its own unique ESN. When a cellular subscriber attempts to place or receive a call, the MTSO first checks its data base to determine if the caller is a valid subscriber. If so, the call is processed. Electronic serial numbers also allow cellular carriers to maintain a list of stolen units, bad credit risks, and other parties to whom they do not want to provide service.

ROAMING

In order for cellular telephones to be commercially successful, they needed to be able to operate in any city with cellular service. Most carriers have reached agreements with each other to provide their customers with service when they travel to other cities. When a cellular customer travels to another city and uses a different carrier's service, the process is called *roaming*. While in the foreign city (a foreign city is a city other than the subscriber's city of registration), the cellular carrier collects billing information on the "roamer" and sends it back to the customer's home carrier. The charges for tolls and air time then appear on the cellular customer's bill at the end of the month.

In geographic areas where there is a great deal of "roamer" traffic, a technique called *dynamic verification* is used. The carriers exchange ESN data to ensure the "roamers" are valid subscribers. This verification technique has saved carriers thousands of dollars in fraudulent charges.

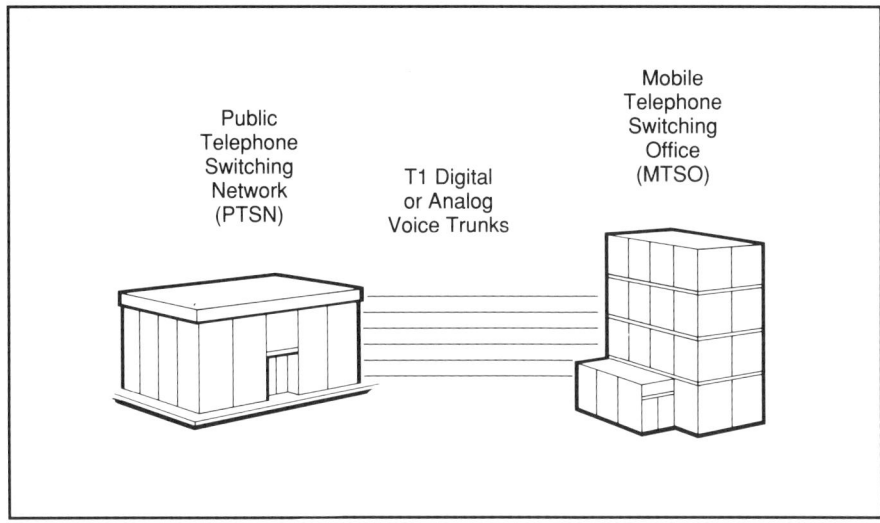

FIGURE 2.8 INTERFACE OF MTSO WITH PTSN
The MTSO is connected to the PTSN by either T1 digital or analog voice trunks.

LANDLINE INTERFACES

The MTSO is interfaced with the public telephone switching network via either analog or digital voice trunks (see figure 2.8). T1 digital trunks are usually the connection of choice since they offer an economical way of connecting a large number of telephone trunks to a landline switching office. Most modern cellular MTSOs offer direct T1 connections to the landline switching office.

These trunks are divided into specific trunk groups. Calls originated by the cellular customer that are terminated in a POTS are processed in the following manner. The MTSO seizes a trunk in the group that will result in the least cost to the calling party. A least-cost routing software module in the MTSO contains this information. The digits are then out-pulsed to the land network that will process the call. An incoming call is processed by the switch collecting signaling digit from a seized trunk. The MTSO then signals the mobile terminal via a cell site.

MTSO DESCRIPTION

Functions

The purpose of an MTSO is to provide the switching functions required for cellular mobile telephone operations and to interface with the public telephone switching network. Most modern digital MTSOs employ time division multiplexing (TDM) and pulse code modulation (PCM) techniques. An MTSO provides the basic telephony switching functions required for cellular operation and for overall coordination and control of each cell site (see figure 2.9). These functions include the following:

1. SWITCHING OF VOICE CHANNELS TO ACCOMPLISH END-TO-END CONVERSATIONS FOR LAND-TO-MOBILE, MOBILE-TO-LAND, AND MOBILE-TO-MOBILE CALLS. This switching facilitates the handoff process and allows continuous conversations as mobile terminals travel from cell site to cell site. It also provides for dynamic channel allocation, which improves the transient traffic-handling characteristics of the system.

2. CONTROL AND DETECTION SIGNALING TO AND FROM THE PUBLIC TELEPHONE SWITCHING NETWORK. This provides for the control and direction of land-to-mobile and mobile-to-land conversations.

3. CONTROL AND COORDINATION OF INFORMATION AND SUPERVISION SIGNALING TO MOBILE TERMINALS. All communication with mobile terminals is controlled from the MTSO.

4. CONTROL AND COORDINATION OF CALL-PROCESSING ACTIVITIES FOR THE MTSO AND CELL SITES.

5. CONTROL OF ON-LINE ADMINISTRATION, MAINTENANCE, AND DIAGNOSTIC ACTIVITIES. Cell site and MTSO status and operational condition are all controlled by the MTSO.

6. CONTROL OF VERTICAL FEATURES AND OTHER CELLULAR SUBSCRIBER SERVICES. Call forwarding, call waiting, conferencing, and access to special services such as stock quotes and traffic conditions are controlled by the MTSO programming. The vertical feature modules of the switch programming are usually designed so that additional features can be added as they become available.

7. CONTROL OF DATA LINKS BETWEEN THE MTSO AND CELL SITES. These links can be anything from copper landlines to microwave stations and fiber optical cable. Copper leased lines offer the advantage of being easy and inexpensive to install. The cellular carrier (wire or nonwire) simply contacts the landline service provider and orders the lines. The responsibility for operation and maintenance belongs to the landline

FIGURE 2.9 AN MTSO

service provider. The cellular service provider does have to pay a monthly charge to the landline carrier for the use of the copper lines.

An alternative for the cellular service provider is to install its own lines. It is often economically attractive to install microwave stations and bypass the landline carrier. In many cases, the installation of a microwave system will pay for itself within two years. Fiber optical cables can be economically attractive in certain situations where conduit or other protective enclosures are already available. A single fiber optical cable has the theoretical capability of carrying over 10 million conversations.

8. CONTROL OF THE SYSTEM DATA BASE. This data base contains subscriber information such as telephone numbers, electronic serial numbers, etc. The data base also contains traffic statistics and other information relevant to the system operator.

Hardware

Most modern cellular telephone systems are configured in the same basic manner. A control chassis houses a processor, memory, timing circuits, the switching matrix, and interface modules. The computer holds and executes all of these stored program processes. The total operation of the system depends upon the proper operation of the processor.

A telephony chassis holds the line, trunk, data links, and other interface modules. The interface of all the remote sites to the public telephone switching network is facilitated through this chassis. Power distribution equipment provides AC interfaces, power protection, battery chargers, and DC power distribution. To ensure high levels of reliability, major components are redundant and on hot stand-by. Each control half is capable of controlling the entire switch, although only one half is active at any given time. The inactive portion of the switch is constantly updated so that in the event of a failure the system would switch control to the stand-by side. Connection of the two systems is usually accomplished through the use of a redundant data bus. Noncritical items (portions of the switch that are not required to process calls) may be nonredundant.

System Software

MTSO system software must provide for control of all telephone switching functions. These functions include call processing, maintenance, administration, and man/machine interface. The major portion of the cost of any MTSO is the software development. The software is usually written in a high-level language to simplify the coding process. Modular design is a must, since dozens of software people will be responsible for developing the operating system. Modular design also allows for easier changes as the customers require them.

CELL-SITE SYSTEM DESCRIPTION

A cell site is the portion of a cellular mobile telephone switching system that communicates with the mobile terminal and interfaces it with the MTSO (see figure 2.10). Its function is to allow geographical distribution of RF coverage to the mobile terminal. Most cell sites are self-contained structures, remotely located and connected to the MTSO via any number of communications methods. These methods can include copper landlines, microwave, and fiber optical cables. Most modern cell sites include the following:

1. FACILITIES TO MONITOR AND CONTROL MOBILE TERMINALS. By properly monitoring mobile terminal status, a cell site (with the aid of the MTSO) can determine when to handoff a conversation and when to release the channel (at the termination of a call).

2. MOBILE POWER-LEVEL CONTROL. By reducing the power level of a mobile terminal in situations that permit such an action, co-channel interference can be significantly reduced.

3. CELL-SITE CHANNEL CONTROL. Voice channel assignment is usually controlled at the cell site. By distributing the processor demands to where they are needed, MTSO processor demands are reduced, and the overall system traffic-handling capacity is improved.

4. TRANSMISSION, DETECTION, AND REPORTING OF MOBILE TERMINAL SATs (SUPERVISORY AUDIO TONES). These tones allow the MTSO and mobile terminal to determine which cell site a mobile unit is communicating with, since the same channel can be reused a number of times within the same CGSA. Three SAT tones are available for use by the MTSO.

5. TRANSMISSION OF SUPERVISION, ADDRESS, AND INFORMATION SIGNALING TO MOBILE TERMINALS.

6. PAGING-CHANNEL MESSAGE FORMATTING AND TRANSMISSION CONTROL.

24 THEORY AND OPERATION OF A CELLULAR SYSTEM

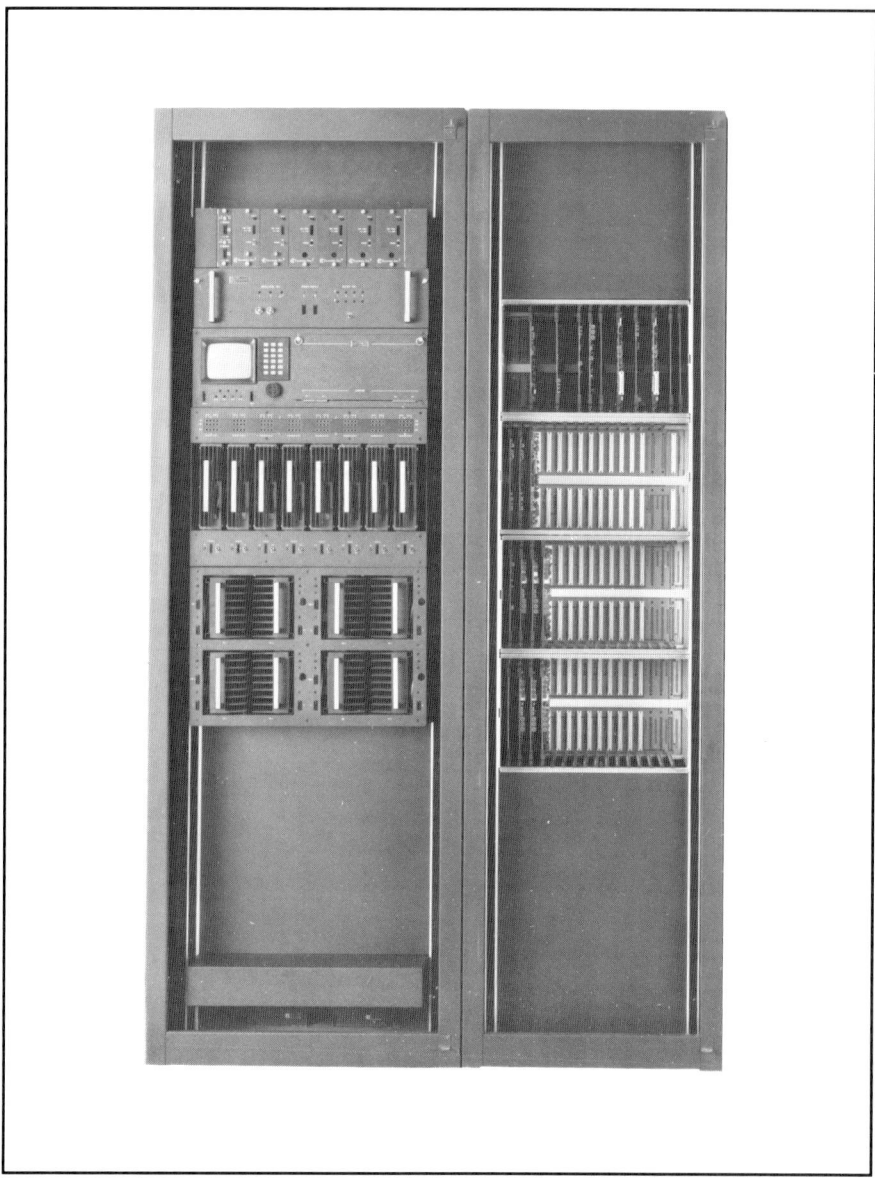

FIGURE 2.10 A CELL SITE

7. LOCAL ALARM AND STATUS MONITORING AND REPORTING.
8. DATA AND VOICE-LINK CONTROL.
9. PROCESSING AND CONDITIONING OF VOICE AUDIO FOR TRANSMISSION TO AND RECEPTION FROM MOBILE TERMINALS. One of the important processes performed on the voice signal is *com-*

panding. This technique compresses speech during transmission and expands it after reception by using a variable-gain linear amplifier. The result is a significantly improved signal-to-noise ratio of the voice signal. This procedure is similar to the one used by Dolby™ and DBX™ for reducing the signal-to-noise ratio of high-fidelity recordings. The process of syllabic amplitude companding suppresses intersyllabic noise and increases the RMS deviation of the FM carrier. To work effectively, the gain characteristics of the transmitter must be matched to the receiver. If a proper match is not achieved, the signal-to-noise ratio is degraded.

A cell site usually contains the following equipment:

1. VOICE-CHANNEL EQUIPMENT. This is the radio equipment that communicates with the mobile terminal. The greater the traffic demand, the more voice channels are required at a cell site. Each conversation requires one voice channel.

2. CONTROL-CHANNEL EQUIPMENT. The control-channel equipment provides for paging and access channels. Control channels can be increased as traffic demand increases. Since a control channel is used only during the seizure portion of the conversation, one control channel will support a number of conversations. Each cell site usually contains one or two control channels.

3. LOCATING-CHANNEL EQUIPMENT. This equipment is used to determine the approximate physical location of a mobile terminal.

4. POWER-DISTRIBUTION EQUIPMENT. This equipment provides the proper power to the various components of a cell site. It also ensures protection in the event of a circuit failure.

5. TRUNK-INTERFACE EQUIPMENT. This equipment furnishes the distribution of voice signals to the MTSO.

6. DATA-LINK INTERFACE. This interface connects a cell site to the MTSO so that pertinent data can be transferred between the MTSO and cell site.

7. ANTENNA EQUIPMENT CONSISTING OF CABLING, TOWER, AND ANTENNAS. Since high-frequency signals tend to propagate along the surface of a conductor, very thick cabling must be used to get the RF signal from the cell site to the antenna.

Often, 7/8-inch or 1-inch heliax cable is the choice of carriers for this task. Antennas are mounted on existing towers or ones erected especially for this task. In high-density urban areas, existing towers often dictate the location of a cell site, since space to locate a tower is at a premium.

Space-diversity receive antennas consist of two antennas with a voting device to select the antenna with the best signal. They offer a significant reception advantage over a single antenna, because the chance is small that two antennas in different physical locations would both be in poor-signal areas. Other diversity methods, such as frequency or time diversity, are available. Frequency diversity is achieved by transmitting the same information on more than one channel. While it is effective, it is not spectrum-efficient. Time diversity involves transmitting the same information more than once. However, this is not practical in a real-time environment.

8. RF COMBINERS. These allow several channels to be transmitted from one antenna. This is done for economic reasons, since combiners induce loss.

CHAPTER 3

Test Equipment and Tools

I worked for a cellular installation company in a rather "tough" section of New York City. The owners felt that by locating very close to the Lincoln Tunnel in Manhattan, they could attract the commuter population from New Jersey, along with the upscale Manhattan crowd. Unfortunately, this put our shop right in the middle of New York's infamous "Hell's Kitchen."

One weekend, the shop was broken into. Once inside the building, the burglars attempted to "crack" the safe. Fortunately for us, they tried this with a hammer and chisel. After they discovered the safe had 6-inch solid concrete walls, the job was abandoned, and the potential thieves left with nothing. The moral of the story: Use the right tools for the job!

Installing cellular phones is like most other jobs—with the right test equipment and tools, it can be easy, enjoyable, and profitable. Without them, the task becomes frustrating and difficult. Selecting the proper test equipment and tools is the first important step in setting up a service and installation facility. Without the tools, it is impossible to install the phones. Without the test equipment, it is impossible to troubleshoot defective installations or equipment. These are the two basic services that a cellular installation facility provides for its customers.

Some cellular installation facilities feel that it is better for installers to provide their own tools. In my opinion, this practice is a disservice to the facilities and to their customers. When service and installation facilities purchase the tools, they can be sure that the proper equipment is available for each installation. If there is concern about theft, the company can make each installer financially responsible for the equipment in his/her cabinet.

TEST EQUIPMENT

Cellular Test Center

The most important piece of equipment you can purchase for a cellular installation facility is a cellular test center (see figure 3.1). A cellular test center is an automated measuring instrument that aids a technician while installing and troubleshooting cellular phones. A good unit will automatically and quickly perform trouble diagnosis and performance tests. What makes the cellular test center such a valuable tool is its operation as a stored-program device. A stored-program device means the test center has a built-in program that automatically performs a series of tests in a sequence and displays the results for the technician. A properly trained technician can interpret the results and quickly determine if a problem exists or if the cellular phone is working properly.

The test center also performs the functions of many different pieces of test equipment and thus can save the installation facility a significant amount of money. When looking for a cellular test center, you should look for several features.

1. POWER METER. The test center should have a self-contained power meter. The meter should read up to 5 watts (in case the cellular phone is putting out too much power) and should also be easy to read at low power levels. This will allow the service center to measure the cellular phone's maximum power output, in addition to its power output at various power levels.

2. FREQUENCY COUNTER. A frequency counter measures the frequency of the carrier generated by the cellular phone. The frequency of the carrier determines the channel on which the cellular phone will transmit and receive. If the frequency is not correct, the phone will not operate on the right channel. The result is a phone that won't make calls.

3. SIGNAL GENERATOR. A signal generator is used during a number of automatic tests that the test center should be able to

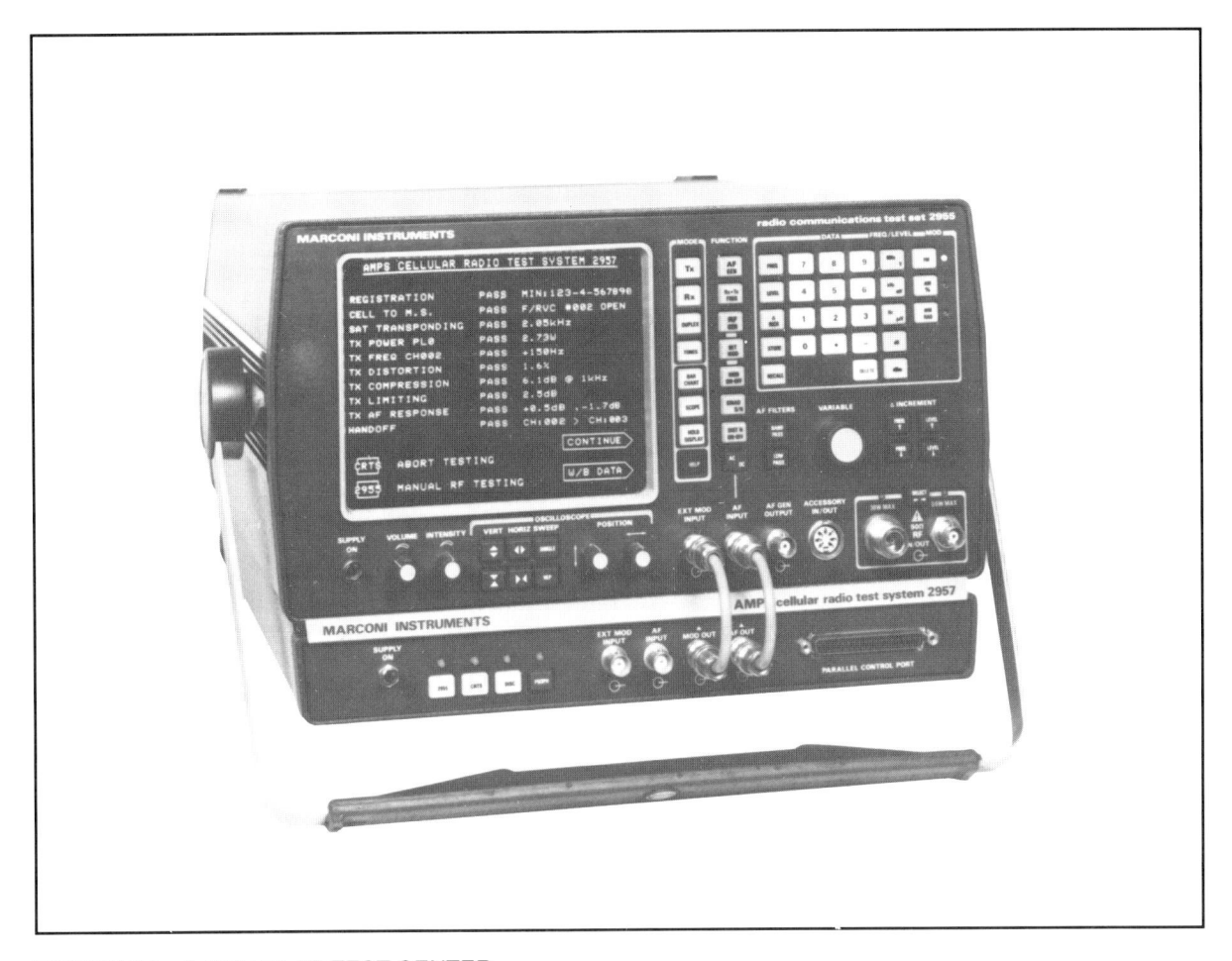

FIGURE 3.1 A CELLULAR TEST CENTER

execute. This device generates signals on channels that are selected by the technician. This piece of equipment is very useful in performing tests on receivers, since it can generate signals of predetermined strength and accuracy.

4. OSCILLOSCOPE. A good oscilloscope allows the service technician to look at the modulated carrier in the time domain. This means that it displays a "picture" of the signal in the form of amplitude (or signal strength) versus time (see figure 3.2). By looking at this display, the technician can determine if the cellular phone's transmitter, modulator, and audio sections are working properly.

5. SPECTRUM ANALYZER. This device displays a "picture" on a CRT (cathode ray tube) of amplitude versus frequency (see

30 TEST EQUIPMENT AND TOOLS

figure 3.3). This equipment is also useful in determining proper operation of the cellular phone's transmitter section. It displays the carrier and sidebands in a clear, easy-to-understand manner.

6. AUTOTEST MODE. The autotest mode should automatically test the specifications listed below to ensure that they are all within AMPS limits (which are explained later in this chapter). It is very desirable for the test center to be able to do these tests automatically, since manual execution can be rather time-consuming. By using the autotest mode of a cellular test center, the service facility can test every phone when it arrives from the manufacturer.

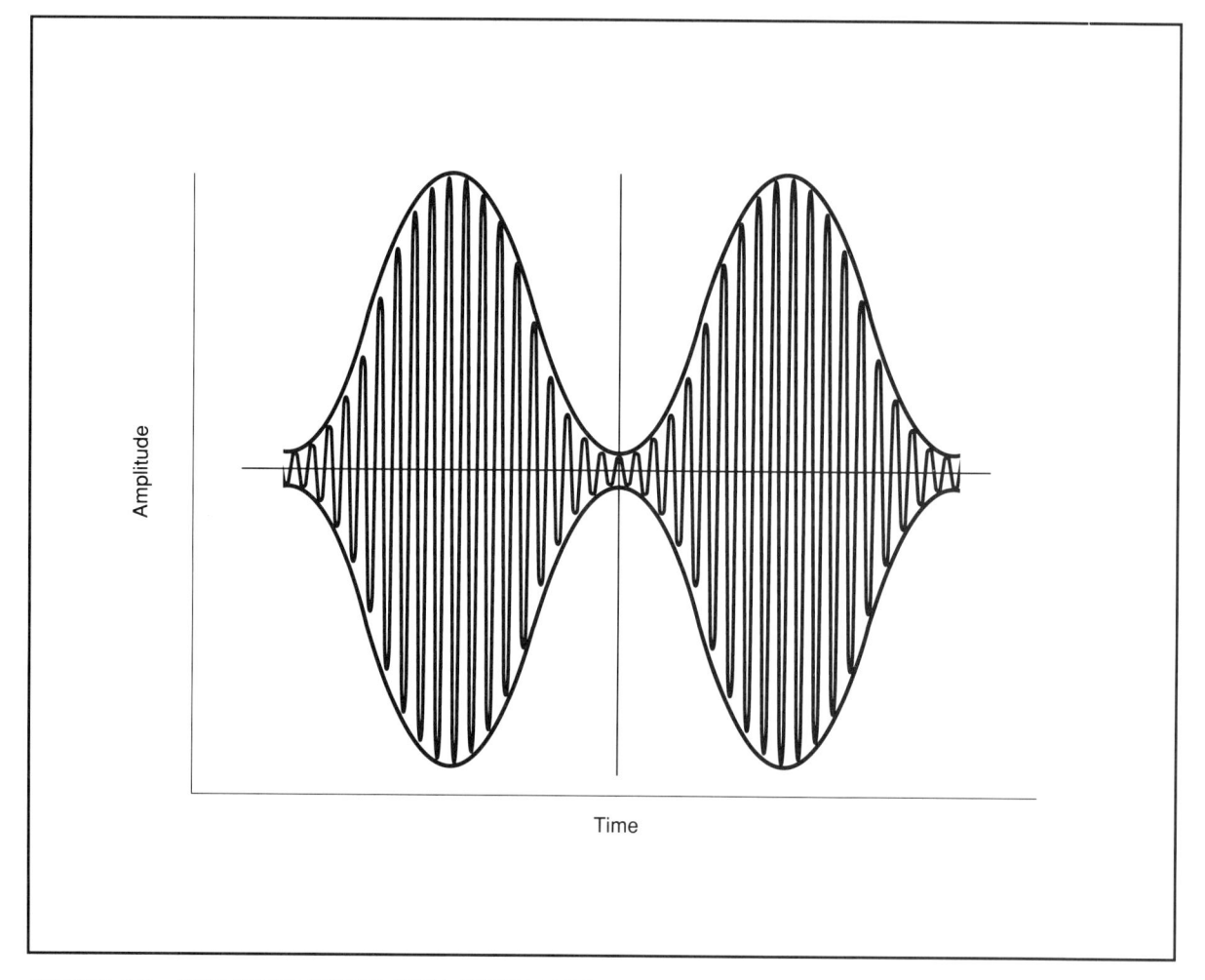

FIGURE 3.2 AMPLITUDE MODULATED SIGNAL DISPLAYED ON AN OSCILLOSCOPE

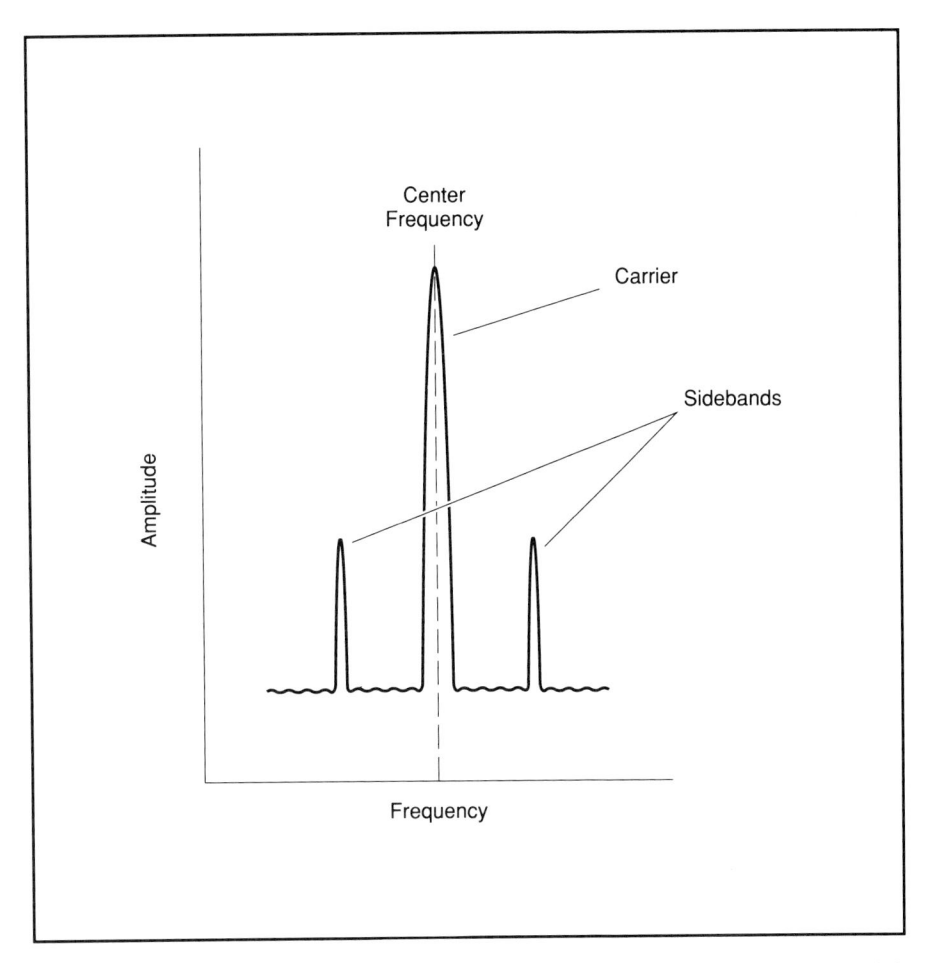

FIGURE 3.3 AMPLITUDE MODULATED SIGNAL DISPLAYED ON A SPECTRUM ANALYZER

The autotest should be able to check all of the functions and the performance of a cellular phone. If a phone passes the autotest on a good test center, it should work properly in a cellular system. This reduces the chance of installing a defective phone in a customer's car and then having to remove it.

AMPS Specifications:

Transmitter power = ±2 dbm from specified power levels 0–7

SAT deviation = 2 kHz peak ±10 percent

ST deviation = 8 kHz peak ±10 percent

DATA deviation = 8 kHz peak ±10 percent

ST frequency = 10 kHz ±1 Hz

SAT phase = ±20 degrees

Transmitter distortion = 26 db or better

SINAD sensitivity = –116 dbm or better at 12 db SINAD

Receive audio level = 20 dbV ±1 db

The autotest should be able to check the following items:

1. Transmitter output power
2. Transmitter frequency
3. AF (audio frequency) modulation
4. SAT (supervisory audio tones)
5. ST (signaling tones)
6. DATA (composite data formats)
7. DTMF (dual-tone multi-frequency) tones
8. Maximum frequency deviation
9. Transmitter distortion
10. ST frequency
11. SAT phase
12. Receiver sensitivity
13. Receiver distortion
14. Demodulation level
15. RSSI (relative signal-strength indicator)
16. Antenna VSWR (voltage standing wave ratio)

The test center should also be able to perform the following tests in manual mode:

1. MANUAL CHANNEL ASSIGNMENT. The test center should have the ability to turn on a cellular transceiver's transmitter and assign it to a particular voice channel. This allows measurement of a number of important parameters of operation.

2. INITIATION OF MANUAL HANDOFFS. One of the more important functions performed by a cellular phone is a handoff. The test center should have the ability to force the mobile unit to perform manual handoffs at any signal level.

3. TRANSMISSION AND RECEPTION OF CELLULAR PHONE NUMBER, ELECTRONIC SERIAL NUMBER, AND STATION CLASS MARK. This test permits the technician to quickly identify a particular cellu-

lar phone and to make sure the number assignment module (NAM) has been properly programmed.

4. **PAGING AND ASSIGNMENT OF A VOICE CHANNEL.** The test center should be able to page a cellular phone. This means that it can send the data necessary to place a call to a cellular phone. The test center should be able to receive the acknowledgment from the phone and instruct it to retune its transceiver to a voice channel. This process mimics a cell site during the set-up portion of a phone call.

5. **PROCESS A MOBILE-INITIATED CALL.** Without using the cellular system, this test ensures that the mobile phone is able to place a mobile-to-land call.

6. **PROCESS A CELL-SITE-INITIATED CALL.** Without using the cellular system, this test ensures that the mobile phone is able to receive a land-to-mobile call.

Dual-power capability (110 volts AC and 12 volts DC) is handy to have with a cellular test center. This permits the test center to be used on a service bench or operated from a 12-volt supply, such as a car battery, in a field environment (a great advantage when on a remote installation).

You should also make sure the test center manufacturer offers interfaces for all cellular telephone manufacturers. *Be very sure to have this demonstrated before purchasing a test center.* This is important, since most manufacturers have their own proprietary interfaces.

The final feature to look for in a test center is ease of use. The most sophisticated test center serves no use if the technician does not know how to use it. Look for things like menu operation and a manual written in clear, easy-to-understand language. Test centers are rather like cars: The choice is a personal one. Test centers are also very expensive. They range in price from a few thousand dollars to almost twenty thousand dollars. Take your time, and look at as many units as possible. (A "Directory of Manufacturers" is included at the back of this book.) The makers of these test centers will be more than happy to visit your facility and answer all of your questions. Ask the salesperson for a complete demonstration, and have him/her leave a demonstration unit for a few days. After all, you wouldn't buy a car without test driving it first.

Watt Meter

Fortunately, the cellular test center is the most expensive piece of equipment you will need to purchase for a cellular installation and service facility. A watt meter, while much less expensive, is just as important to a quality installation. This device serves two functions.

The first function is to measure the amount of power being delivered to the antenna by the transmitter in the cellular phone. The power output of the phone should be about 3 watts for a mobile phone (except for some class II mobiles) and 0.6 watts for a hand-held portable phone. It is important to remember that some cellular service providers in larger markets are using a technique called *dynamic power-level control* (see Appendix A, "Glossary"). The purpose of this technique was explained in Chapter 2. It is important to mention here because the cellular service provider may instruct (via a set of data commands) the cellular phone to turn its power output down. If this happens, the watt meter will read 3 watts for less than a second and then will drop to something less than that. The power output will depend upon the power level that the cellular unit was set to by the cellular service provider. If you do not know this, you may think the unit is putting out less than 3 watts and thus attempt to repair the phone. You will spend a lot of time looking for a fault that is not there. You should also remember that the service provider (carrier) may change the power level a number of times during the conversation. If this happens, use the cellular test center to manually turn the transmitter on at full power and measure the power output with the watt meter.

Once you are convinced the cellular phone is transmitting (or capable of transmitting) at 3 watts, the watt meter should be used to measure the power reflected back from the antenna. Determining the amount of reflected power helps the installer keep it to a minimum. In Chapter 5, we will discuss reflected power and why it is important.

NAM Programming

Every cellular phone contains a *number assignment module,* or NAM. The NAM is programmed by the installation facility and contains several pieces of subscriber information. Each carrier has specific information that needs to be programmed into the subscriber's cellular phone. The NAM contains the following information:

1. SIDH. A 15-bit field in the NAM designates the system identification of the home system. (This field is binary, meaning that it is a string of 1s and 0s.) Each cellular carrier has a unique SIDH for each system it operates. For example, the New York City wireline carrier has an SIDH of 00022. The nonwireline carrier in Buffalo, New York, has an SIDH of 00003. All cellular phones registered in the same system have the same SIDH programmed into their NAMs. The first bit of the SIDH represents the preferred system flag. Bits 5 and 6 of byte 0 are international code bits. The U.S. code is 00, Canada is 10, and Australia is 01.

2. LOCAL-USE FLAG (OR MARK). This tells the cellular unit if it must preregister with the system. Preregistration with the system means that the cellular phone must transmit its parameters (ESN, cellular phone number, station class mark, etc.) to the cellular system before a call can be made. Setting the local-use bit to 1 enables this function. The function should always be enabled when a cellular phone is to be used on a U.S. or Canadian system.

3. MIN. The mobile identification number is often referred to as the cellular phone number. It is a 10-digit number that is formatted just like a standard POTS phone number. The format is as follows:

    NPA-NXX-XXXX

    where NPA is the area code, NXX is the exchange, and XXXX is any four numbers. This is one of three pieces of information that will be different in each cellular phone in the same cellular system.

    Some cellular phones have the capability of dual registration. Such units are said to have "dual NAM capability." This means that they have the capability of subscribing to service from two different carriers at the same time. Why would anyone want to subscribe to service from two carriers? A person who frequently travels from New York to Los Angeles might want to have local numbers in each city. Since roaming is so complicated in North America, having a phone number in a frequently visited city makes a lot of sense. Another reason for having two phone numbers in one cellular phone is to make up for differences in coverage between carriers. The wireline carrier might have better coverage in the downtown area, while the nonwireline carrier might have better coverage in the rural areas of the same system. A cellular customer could subscribe to both systems in order to always have good coverage.

A dual NAM unit will *not* scan both systems at the same time. This means that a cellular customer does not have the ability to be "listening" for a call on both systems in the same city at the same time. The reason for this has to do with timing parameters that are too lengthy to discuss here. The real value of dual NAM phones is the capability of having local numbers in two cities.

4. SCM. Station class mark is a 4-bit field. The first bit is a flag that indicates the number of channels the cellular phone will scan. If the bit is set to 1, the phone (if it is equipped to do so) will scan the "new" cellular channels. This gives the phone access to all 832 channels. Older cellular phones not equipped to access the new channels should have this bit set to 0. These units can access only 666 channels.

The second bit determines if the unit will transmit continuously or only when the cellular phone party is speaking. A 0 indicates continuous transmission. Since most systems support only continuous transmission, the bit should almost always be set to 0.

The last two bits indicate the power class. A cellular phone capable of 3-watts power output should have these bits set to 00. A 1.2-watt maximum power unit should have these bits set to 01, and a 0.6-watt unit to 10.

5. IPCH. Initial paging channel is an 11-bit field that instructs the cellular unit on which paging channel to begin looking for information. The units programmed with wireline phone numbers should have an IPCH of 334 and nonwireline numbers, 333. Some NAM programmers automatically set this field when the SIDH is entered, since even-numbered SIDHs indicate wireline and odd-numbered SIDHs indicate nonwireline.

6. ACCOLC. Access overload class is a 4-bit field in the NAM. The intention of this field was to allow the cellular system operator to determine the priority of access in the event of a system overload. Presently, most operators are not using this feature on the system. It is usually set to 00 by the installer. Refer to the instructions from the cellular service provider to be sure.

7. PS. Preferred system is a 1-bit field that instructs the cellular phone to scan the wireline (B) or the nonwireline (A) system. Some NAM programmers automatically set this bit when the SIDH is entered, since even-numbered SIDHs indicate wireline and odd-numbered SIDHs indicate nonwireline.

8. GIM. Group identification mark is a 4-bit field that tells a cellular unit how far to look in the SIDH to determine if it is roaming in a system that may have a roaming agreement with its home carrier.

9. EE. End-to-end signaling is a 1-bit flag that determines if the cellular phone can transmit DTMF tones on a voice channel. This feature must be enabled if the cellular phone will be used with DTMF-controlled devices such as voice mail, alternate long-distance carriers, home answering machines, etc. This bit is almost always set to 1.

10. REP. The repertory memory bit enables the unit's repertory (speed-dialing) memory. The bit should be set to 1.

11. HORN ALERT. This is a 1-bit field that is set to either 1 or 0. When it is set to 1, the feature is activated. If it is set to 0, the function will not operate. When the function is activated, the phone can be left on after the vehicle has been turned off. If the cellular unit is properly equipped with external hardware, the horn will sound when the phone is called. This is a very handy feature for someone who is outside (but near) the car and wants to have use of the car phone. The cellular unit must be equipped with a relay and connected to the vehicle's horn in order for this feature to work. Most installation facilities charge an additional fee for this option.

12. HANDS-FREE. This is activated by setting the hands-free bit to a value of 1. This informs the cellular unit that it has been equipped with the hands-free option. This bit changes the operation of the cellular phone so that the end command is not sent when the hand set is placed back on the hook. This lets the user switch from hand-set to hands-free operation by placing the hand set back on the hook. The bit should be set to 1 for OEM-supplied hands-free equipment. Some aftermarket equipment manufacturers require that the bit be set to 0 for proper operation. To be sure, refer to the manufacturer's hands-free instructions.

NAMs (Number Assignment Modules)

Different manufacturers of cellular phones have different thoughts on NAM programming. Some manufacturers favor the approach of a dealer-removable PROM (programmable read-only memory) that contains the subscriber information (see figure 3.4). These NAMs need to be programmed by a device called a *NAM programmer*. These are manufactured by a number of companies and are designed specifically for the cellular market (see figure 3.5). Most NAM programmers offer menu-driven operation and are easy to use. Dealer-removable PROMS can be installed faster and are less prone to programming errors than EEPROMS. The dealer can program a master NAM and then copy the contents into a removable NAM. Only the phone number and a few options need to be changed with each different cellular telephone.

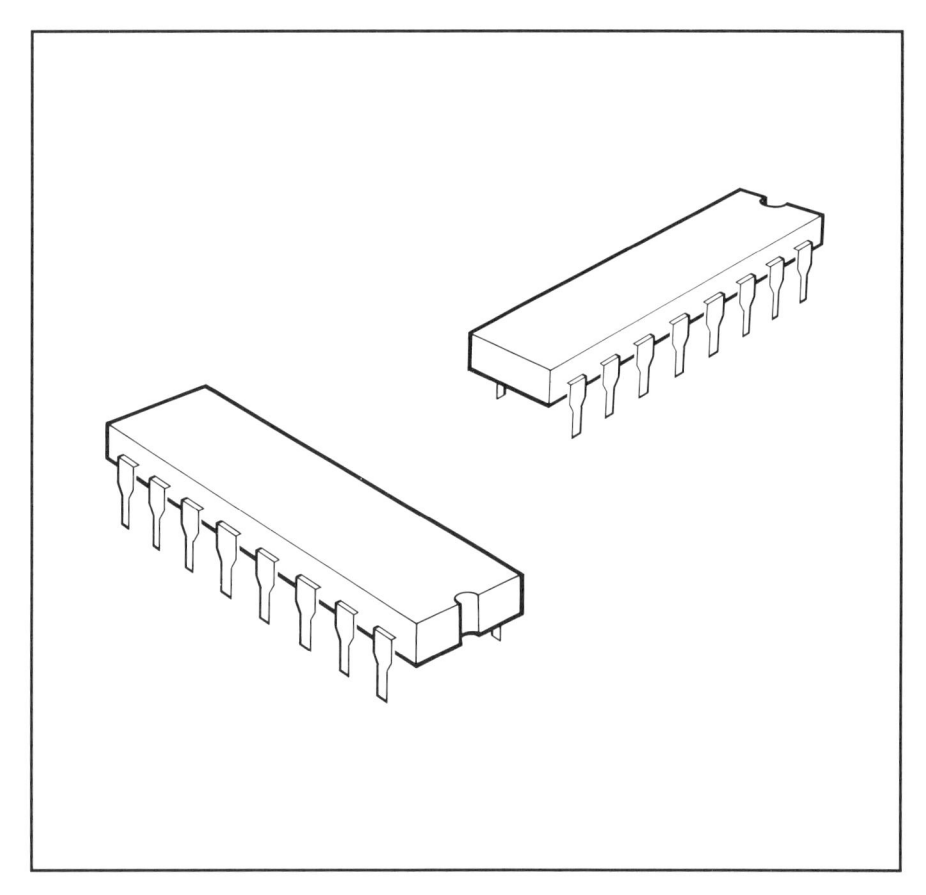

FIGURE 3.4 NAMs

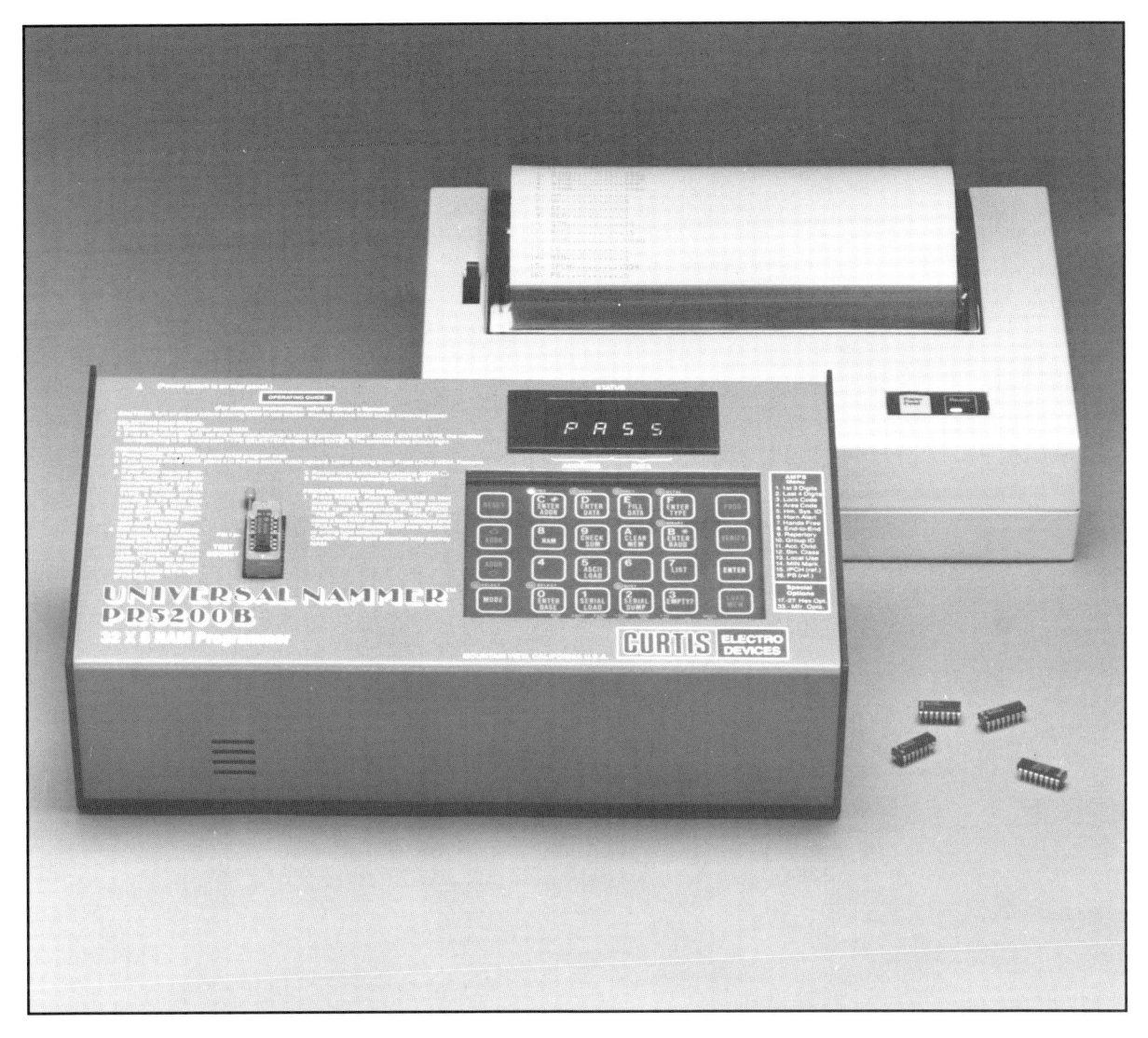

FIGURE 3.5 A NAM PROGRAMMER

Other manufacturers use an internal, nonremovable EEPROM (electronically erasable programmable read-only memory). The EEPROM is usually programmed from the customer's handset or through the use of a special programming handset (see figure 3.6). One of the advantages of this method is low cost for the dealer. However, if a manufacturer requires the use of a special programming handset, it allows them to control who will install their product.

FIGURE 3.6 A SERVICE HANDSET

Power Supply

In order to properly and simply test a cellular phone, a service facility needs a good power supply. This enables the technician to test cellular phones outside of the vehicle. The last thing a technician wants to do is troubleshoot a malfunctioning cellular phone in the trunk of a car that has been used to haul goat droppings. A good bench power supply should deliver well-regulated 12 volts DC at 5 amperes minimum. It is essential for the power supply to be properly regulated so that no AC or RF noise is introduced into the cellular phone while troubleshooting. Proper metering on the power supply is also a good idea. It is very important for the technician to be able to determine the amount of current being drawn by the cellular phone. High current drain

TEST EQUIPMENT AND TOOLS 41

or a large voltage drop are often indications of a defective cellular unit. Remember to properly fuse the DC and AC lines. Fuses are cheaper and easier to replace than cellular phones and power supplies.

Test Light

A test light is used to indicate the presence of 12 volts DC. It consists of a cable with an alligator clip on one end and a plastic handle with a light inside on the other end. The handle has a sharp metal probe that can make a small hole in a cable to detect the presence of DC (see figure 3.7). This simple tool can save an installer a lot of time and trouble. Instead of having to slice or

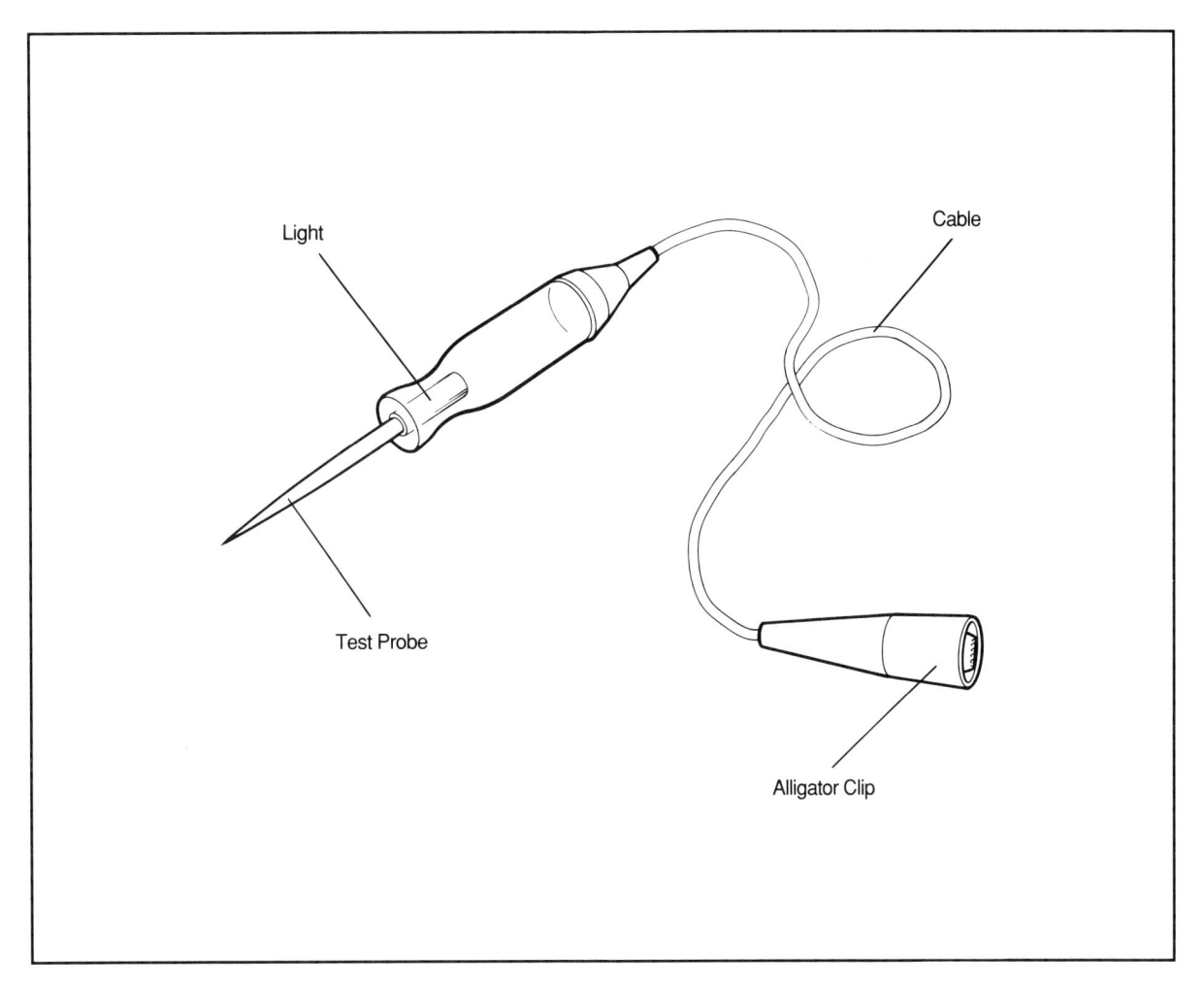

FIGURE 3.7   A TEST LIGHT

disconnect a cable that is suspected of being defective, the installer simply pokes a small hole in the cable and attaches the alligator clip to a piece of metal in the car. If DC is present, the light in the handle will glow. You can buy these at any hardware or auto parts store.

Multimeter

This is another inexpensive piece of equipment that can be a real time-saver. A good multimeter will measure the following:

1. DC and AC voltage
2. DC current
3. Resistance (ohms)

Any electronic supply shop will stock these meters, and a good quality one will usually sell for under $100.

TOOLS

A good selection of hand tools is a requirement for a cellular installation facility. Good tools are an investment that will provide returns. An installer spending time looking for a tool, or trying to figure out how to improvise because the correct tool isn't around, is an installer who is losing money for the service facility. Below is a list of recommended tools. When shopping for these tools, look for a supplier that offers a lifetime guarantee (several manufacturers do this). Don't be afraid to buy retail—it's a great way to get a good buy if you watch for sales. After reading this section, you may think I own stock in a tool manufacturer. I don't! I just know what happens when the right tools are not around when you need them.

1. FLAT-BLADE AND PHILLIPS SCREWDRIVER SET. If you don't have a good screwdriver set around, there is not much you can do in the way of installations. Buy an assortment that contains at least five different sizes of both flat-blade and Phillips screwdrivers. Don't buy cheap ones. Inexpensive screwdrivers will break, or worse yet, strip a screw. It is very frustrating to have to drill out a stripped screw, tap the hole, and replace the screw. It's a process that takes at least half an hour. An extra $10 invested in a top-quality screwdriver set can save an installer a lot of frustration.

2. METRIC AND SAE SOCKET SET. Unfortunately, American car manufacturers have chosen to continue using SAE (feet, inches, pounds, etc.) instead of metric (meters, millimeters, kilograms, etc.) like car manufacturers in the rest of the world. This leaves the installation facility with the unenviable task of purchasing tools to work on both types of vehicles. Most foreign vehicles use metric tooling, while American cars use SAE. Once again, don't try to get by with only one version of sockets. Some SAE sockets will fit metric nuts and vice versa, but most won't. Using the wrong socket on a nut is a great way to strip it. And if you are thinking of just buying a pair of pliers, an adjustable wrench, or vise grips, don't bother. If you don't believe me, watch the expression on the face of a Mercedes-Benz owner as you go after his car with a pair of vise grips.

3. METRIC AND SAE HEX DRIVER SET. This is a tool that you won't need very often, but when you do, you *really* need it. If an installer needs to remove a hex nut and doesn't have this tool, it's off to the hardware store to buy a set. They are not very expensive, so make the investment in the beginning.

4. HOLE-PUNCH KIT. The only correct way to make a hole in a car to mount a roof-top or elevated-feed antenna is with a hole punch. It makes a very clean, controlled hole in metal. The installer drills a small pilot hole and then uses the hole punch to make a hole in the proper location and of the correct size. About three different sizes of hole punches will take care of most cellular antenna requirements. A drill is not a substitute for a punch. Drills have a nasty habit of drifting. This can result in a hole in the wrong place, or, worse yet, the drill can "hop" out of the pilot hole and scratch the trunk or roof of the car. Even if you haven't priced the cost of BMW paint jobs lately, you can be sure that one paint job is more expensive than a set of hole punches.

5. PLUMBER'S SNAKE. Take a drive down to your local plumbing supply house and buy a couple of plumber's snakes. A plumber's snake is a thin piece of spring steel about 15 feet long (or 3 meters, if you are a metric fan). It is very flexible and strong. An installer can tape a cable to it and "snake" it through small openings between the trunk or engine compartment and the passenger compartment. This simple tool is a real time-saver.

6. **Fender covers.** These are available from automobile supply shops for only a few dollars each. As the name implies, they are used to cover the vehicle fenders while installing the phone. It is a good idea to use them while drilling the hole for the elevated-feed antenna, so that a vehicle's paint does not get scratched by metal shavings. They are also useful while making the electrical connections to the vehicle battery.

7. **Eye protection.** This is something that should be worn by all installers while using power tools. A broken drill bit or metal chip in the eye can cause permanent blindness. A $3 pair of goggles can prevent lawsuits and a lifetime of pain. Buy them and use them.

8. **Silicone bathtub putty (clear).** This putty is sold at most hardware stores for under $3 per tube. To prevent corrosion, it can be applied to the sheet metal screws used to mount transceiver plates. It is also a great way to seal holes left when a transceiver is removed from a vehicle. There is no better way to make a customer angry than to remove his cellular phone (to be installed in another car, of course) and then have the car fill up with water because of holes left in the floor.

9. **Battery-powered screwdriver.** This tool will become your best friend in a short time. It has a small motor that turns the screwdriver blade. The blades are removable, so the tool can be used as both a Phillips and flat-blade screwdriver. If you perform four installations per day at an average of 15 screws per installation, that is 120 screws per day. This tool can save a lot of time and make life a little easier. The batteries are rechargeable and removable, so the tool can still be used if a set of batteries needs to be recharged.

10. **Battery-powered drill.** This type of drill is also handy to have in an installation facility. By eliminating the need for a power cord at the car, there is one less thing to trip over in the shop. Both the screwdriver and drill are available at most hardware stores.

11. **Drill bits.** A drill is not very useful without drill bits. Don't buy cheap drill bits because they won't stay sharp. Dull bits take longer to drill holes in cars. They are also more likely to "hop" out of the pilot spot and scratch up the fender or trunk. Good drill bits are not that much more expensive than cheap

ones, and since these bits are used every day, it makes sense to buy good ones.

12. HOLE PUNCH. A hole punch is another one of those tools that costs about $3 and can save the installer a great deal of time and aggravation. Its use will be explained in Chapter 6, "Vehicle Installations."

13. SHOP LIGHT. A good shop light to illuminate those hard-to-see spots in a vehicle is a must.

14. NEEDLE-NOSE PLIERS. These are useful for handling connectors and various other applications. You should purchase several different types.

15. GOOD HARDWARE ASSORTMENT. This can be purchased from a local hardware store and should include an assortment of screws, nuts, bolts, washers, and lock washers.

16. DENTAL MIRROR. This is useful in inspecting hard-to-see areas.

17. WHEEL RAMPS. A set of wheel ramps is a must if you need to get under the car, although this is not usual during an installation. A car jack should never be used to raise the car. Jacks can slip, and having a car fall on you will ruin your whole day.

18. BATTERY CHARGER. Dead car batteries result from leaving a car door open for too long (the dome light draws quite a bit of current) or leaving the vehicle lights on. The only way to recharge a battery is to "jump start" it or to use a battery charger. Using a battery charger is the better method, since it not only gets the vehicle going, but it also recharges the battery. When a car is "jump started," the battery is recharged only while the vehicle is running. If the customer lives just ten minutes away from the service facility, he or she will be unpleasantly surprised to find a dead battery the next morning.

19. SHOP VACUUM. The customer's car should always leave the shop cleaner than it arrived. Be sure to buy a good vacuum that will remove any pieces of cable or other scraps from the car floor. This will make a good impression on the customer. Remember—about 80 percent of your business will come from referrals.

20. SPOT REMOVER. Your best installer has just finished with a Mercedes-Benz. The installation looks great—except for the grease spot on the carpet. Instead of having to explain to the customer how sorry you are, a little spot remover spray will take care of the problem. It can be purchased from any hardware store.

21. SOLDERLESS (OR CRIMP-ON) CONNECTORS. You should have a variety of these available to speed up the job of installing a cellular phone. These connectors are crimped onto a cable, allowing two cables to be joined together. Because these connectors are solderless, they are quick and reliable. No need to worry about cold solder joints or burned carpets from dropped soldering irons. The connectors are sold in most electronic specialty houses or are available from jobbers (who will certainly call upon your new business). Be sure to purchase the proper crimp tool from the manufacturer of the connectors. Don't use pliers—they don't work!

22. VISE. A good vise is a handy piece of equipment to have in your shop. It is used for securely holding circuit boards, connectors, etc. so that they can be worked on easily. Be sure the vise has rubber pads on the face of the grips so that anything delicate will not be damaged.

23. ELECTRIC GRINDER. This machine can be very handy when doing any custom work in a vehicle. Mounting hardware and other equipment may require some customization when doing more sophisticated work. A "Dremil" tool can also be useful.

24. ANGLE DRILL ATTACHMENT. This device allows holes to be drilled in tight areas where a drill will not fit. It's a tool that you won't need very often, but when you do, you'll *really* need it.

25. HOT GLUE GUN. This is an indispensible tool in the installation bay. It is used to reattach carpet, headliners, molding, and a myriad of other things that can come loose during an installation. Hot glue can be a real lifesaver in a lot of situations.

26. WIRE CUTTERS. Don't buy cheap wire cutters or strippers. They will get dull and frustrate the user. It makes sense to buy one good pair and be done with it.

27. TORX BITS. Some phones use "torx" screws to hold the panels on the transceiver. This is done to discourage end users from opening up the phone and attempting to repair it themselves. If you don't have a good torx set, you won't be able to repair the phone either. Very embarrassing.

28. CRIMPERS. In the antenna section of this book, I stress the importance of having the right crimping tool for antenna connectors. Be sure to follow the manufacturer's recommendation when selecting a crimper for antenna connectors.

29. HEAT GUN. If you have ever had to splice a cable and tried to use electrical tape after soldering the wires, you know that it doesn't work. The only way to do this in a professional manner is to use heat-shrink tubing, which is made out of plastic that shrinks when exposed to heat. The tubing is slipped over a soldered joint (after the connection has been soldered) and is then heated with the heat gun; it shrinks to electrically insulate and protect the wire. Buy several sizes of tubing for different types of repairs.

30. TRANSCEIVER KEYS. Some manufacturers use keys to lock their transceivers in their mounting trays for security reasons. It is a good idea to have an assortment of keys on hand. It is a lot easier to work on a phone once it has been removed from the trunk.

FACILITIES

It happens almost every Sunday. Someone advertises on-site installations for cellular phones. Would you want a doctor to remove your appendix in your living room? I doubt it. None of the controls in a shop are available when the installation is done at the customer's site. While on-site installations may appear to offer the customer a bargain, they really expose the customer to a number of risks. What happens if it rains when the installation is only partially complete? Will the installer have the right tools and parts? What happens when the cellular phone breaks? On-site installations are a bad idea.

The installation area does not need to be expensive or lavish. In figure 3.8, a typical one-bay installation facility is shown. A lockable tool chest should either be located on the wall or be mobile. I favor the mobile tool chest, since the tools can be moved closer to the installer. The installation area should have adequate

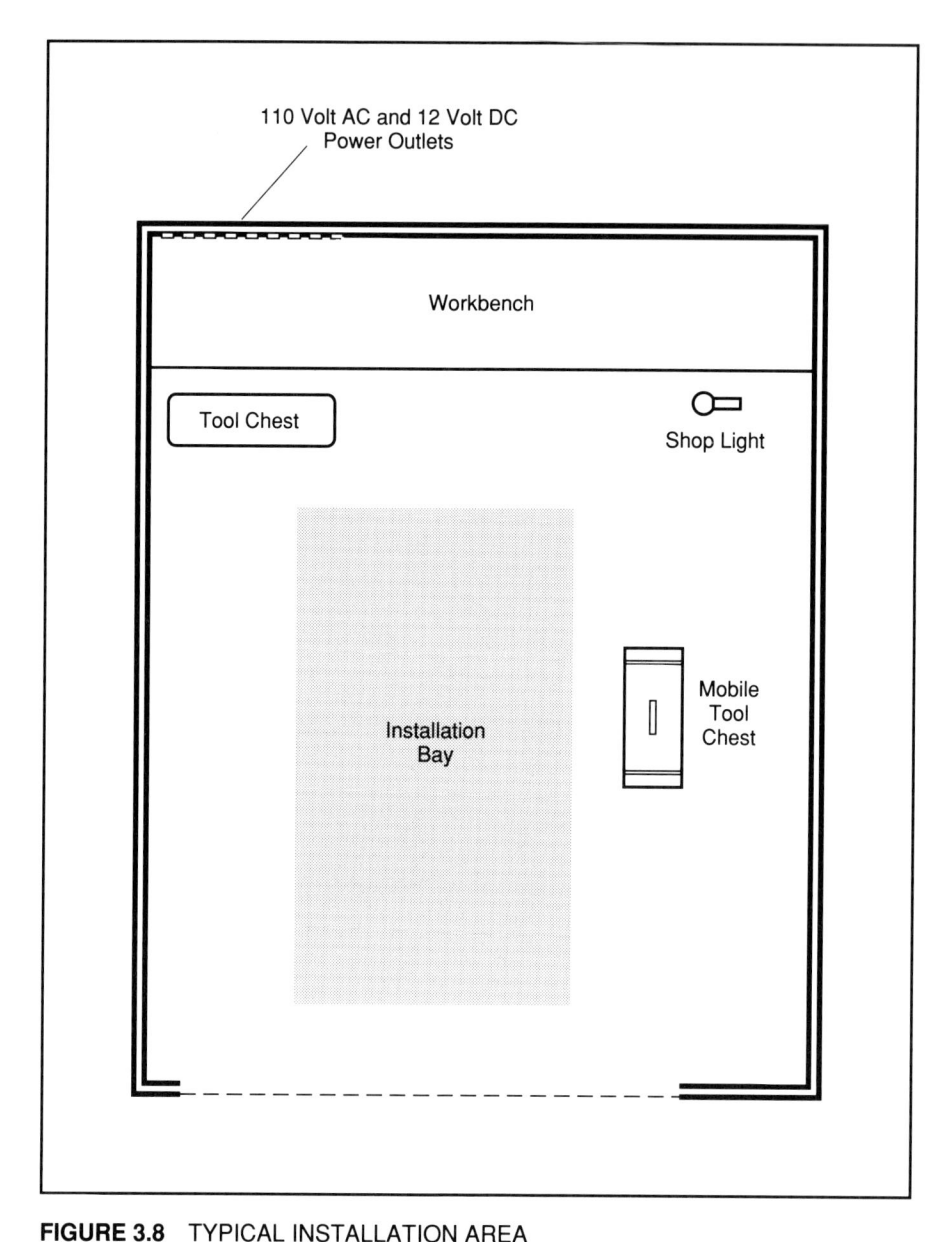

FIGURE 3.8 TYPICAL INSTALLATION AREA
An installation area should be laid out to provide easy access to both the vehicle and the tools.

light and be properly heated and cooled. An installer who can't see or who is freezing is more likely to make a mistake than one who works in a good environment. The shop light should be mounted on a retractable spool so that it will be neatly stored when not in use. These are available on reels at local electronic supply shops.

A workbench should be located somewhere near the installation bay so that items or disassembled pieces of the car can be placed safely away. The workbench should also have a number of power outlets to supply both 110 volts AC and 12 volts DC. Power tools can then be charged and operated without stringing extension cords all over the shop.

I once saw a sign in a service area that read as follows:

Labor Rates

$30 per hour

$40 per hour if you want to watch

$50 per hour if you want to help

The sign made a good point. It is very disconcerting to have someone look over your shoulder while working on an installation. A good way to be sure this doesn't happen is to post signs stating that your insurance does not cover nonemployees in the work area. It is also a good idea to have a waiting area, in case a customer decides to wait for the installation to be completed. Figure 3.9 (on the following page) shows a typical waiting area. It has a few spare desks with chairs and telephones (with long-distance restriction). This is not expensive to set up and will allow your customers to get a little work done while they wait.

A display can also be located in the waiting area so that customers will have a chance to look at optional equipment (or maybe another phone for a business partner or spouse) while they are waiting. Most major cellular phone manufacturers sell point-of-purchase displays for a nominal fee. They are usually professional in appearance and give customers a chance to acquaint themselves with their new purchase. Some manufacturers also offer videotapes that explain the operation of their products. The customer's waiting time can be put to good use by viewing the tapes. This saves the installers time later when they are "checking out the customer." Many variations of this simple floor plan are possible. The goal of any customer waiting area should be to make the customer as comfortable as possible—away from the installation area.

COMPUTER EQUIPMENT

Computers may be one of the most misused tools of the 1980s. Everyone knows that they should buy one, but very few people know what to do with it once purchased. What makes a computer

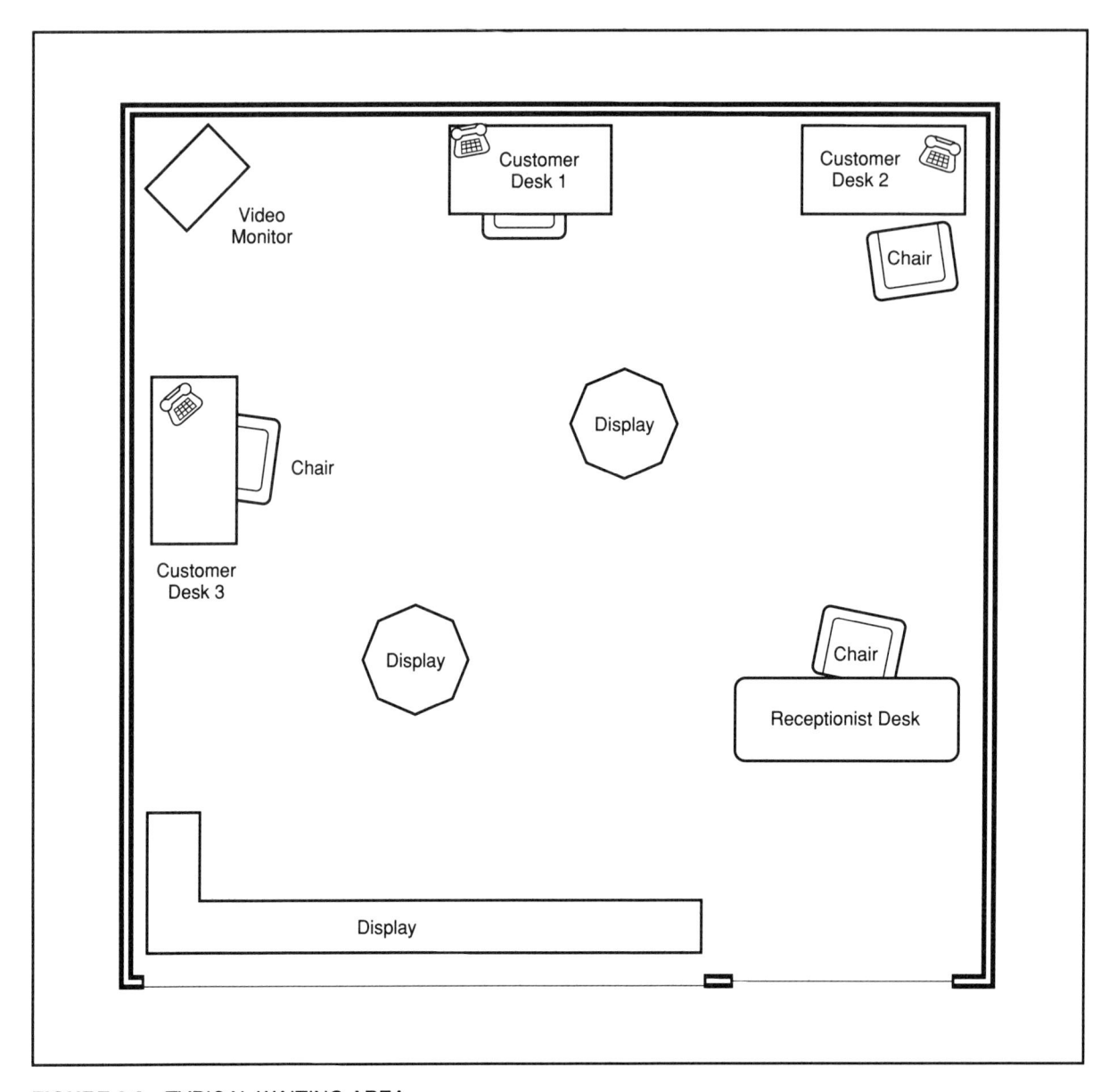

FIGURE 3.9 TYPICAL WAITING AREA
A waiting area provides a convenient place for customers to wait or work while an installation is being done.

system attractive to an installation facility is its ability to store, retrieve, and process records. Keeping accurate records can make the difference between success and failure. The service manager needs to have rapid access to the installation records and a computer is perfect for this task.

This is one of the few places in this book where I will recommend specific equipment. I strongly suggest an MS-DOS, IBM™ PC-compatible computer. It should contain 640K of RAM (random access memory), dual floppy disk drives, and at least 20 Mbytes of hard disk space. The microprocessor should be an Intel 80286 processor. A Hercules™-compatible monochrome graphics card with a good quality monochrome monitor is the best choice.

A monitor with an amber color screen is easiest on the eyes. The computer should also be equipped with a printer port and a 1200 baud modem. A fast dot-matrix printer with letter quality capability is highly recommended. If you watch for sales, the entire system can be purchased for under $3,000.

Now, having purchased this equipment, what can you do with it? The most important application is recordkeeping. By maintaining accurate records of service, installations, and sales, the computer becomes a powerful marketing and administrative tool. There are several software packages that can be purchased to create a file system. These software packages are called data base languages. Some are easy to learn; others are more complicated. Unfortunately, the more complicated ones are usually the most powerful. If you know what kind of records you want to keep, a software designer can usually be hired to write the software, using whatever data base you select. Dbase, Rbase, DataEase, and a few others are good software packages that offer almost any feature you could want.

Some packages already exist that can be modified to fit the individual customer's needs. On the following page is a suggested form that can be used to keep track of your customers.

You may want to add some other fields (areas of information in the record) to meet your own particular needs. Having this information at your fingertips gives your service people a great deal of flexibility. For example, an important customer drives into the shop. The service manager can't remember the customer's name, but a quick check with the computer (search your records by license plate) yields the name. The service manager looks great, and the customer feels important.

The computer also has the capability to sort by any field. This means that you can quickly determine how many phones of a specific type were installed in Chevrolets. Or perhaps how long it takes to install a particular brand of phone in a certain type of car. Even problem customers and "deadbeats" can be tracked with a data base and inexpensive computer.

```
                          INSTALLATION WORK RECORD

COMPANY NAME: _____

CONTACT:  LAST NAME: _____  FIRST NAME: _____

ADDRESS: _____  OFFICE PHONE NO.: _____

CITY: _____  STATE: _____  ZIP: _____

CELLULAR PHONE NO.: (   ) _____  ESN: _____

PHONE MANUFACTURER: _____  MODEL: _____

TRANSCEIVER SERIAL NO.: _____  CONTROL HEAD SERIAL NO.: _____

INSTALLER'S INITIALS: _____  IN-TIME: _____  OUT-TIME: ____

VEHICLE MAKE: _____  MODEL: _____  YEAR: _____

LICENSE PLATE: _____  STATE: _____

SALESPERSON: _____  INSTALLED PRICE: _____

CONTROL HEAD LOCATION: _____

TRANSCEIVER LOCATION: _____
```

A modem allows access to the carrier's MTSO computer. This permits installers themselves to turn on their customers' telephone numbers, query for problems, and generally have almost instant access to information about their customers. They can also electronically exchange information with the carrier, other offices, or other companies. A modem allows access to the electronic world at a very affordable price.

Finally, Lotus 123™ or a similar spreadsheet package can be used to keep track of number activations and your general accounting. Lotus is one of the most powerful packages developed for this purpose and is used by almost every major company in the United States.

Word processing packages are also available. By using a word processor in conjunction with your data base, direct mail pieces can be created from information in your customer file. For example, it would be very easy to send letters to all your customers who own Porsches, offering them a custom-manufactured arm rest that hides the cellular phone from view.

CHAPTER 4

Cellular Telephone Equipment

The next subject we will explore before investigating the world of installation is cellular telephone equipment. This includes not only cellular telephones, but also other peripheral equipment such as interfaces, modems, etc. First, we will discuss the cellular telephones.

CELLULAR TELEPHONES

Cellular telephones are divided into three basic categories:

1. Mobile-only
2. Transportable
3. Hand-held portable

Mobile-Only Cellular Telephones

Mobile-only cellular telephones, as the name implies, are phones that are designed to be installed in a car. They are generally designed to operate at a full 3 watts (with the exception of a few models that are designed to operate at 1.2 watts). The phones range in complexity from simple to sophisticated. In figure 4.1, we see several examples of mobile-only cellular telephones. They usually consist of

1. A transceiver
2. A control head
3. A data cable (to connect the transceiver to the control head)

56 CELLULAR TELEPHONE EQUIPMENT

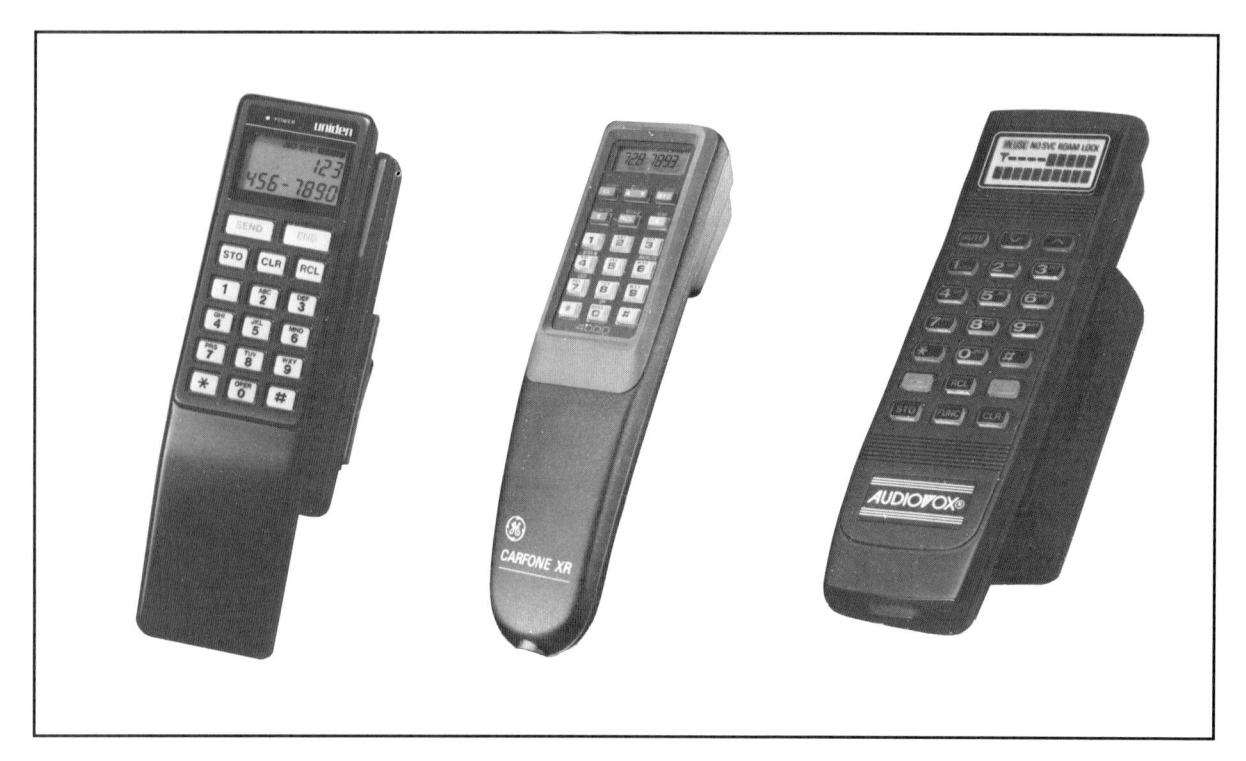

FIGURE 4.1 MOBILE-ONLY CELLULAR TELEPHONES

    4. A power cable
    5. A hands-free microphone (optional)
    6. A hands-free speaker (optional)

Figure 4.2 shows a block diagram of a typical mobile-only cellular telephone. The components that make up a car telephone are described below:

1. TRANSCEIVER. This contains most of the electronics of the mobile-only telephone. The radio, audio, and processors are usually contained in the transceiver.

2. CONTROL HEAD. This contains the man/machine interface, which includes the keypad, speaker/microphone, display, and volume adjustments. The control head is the device with which the customer has the most contact. (Some customers are not even aware that there is a transceiver in the trunk.) An attractively designed control head can blend elegantly with the interior of the vehicle.

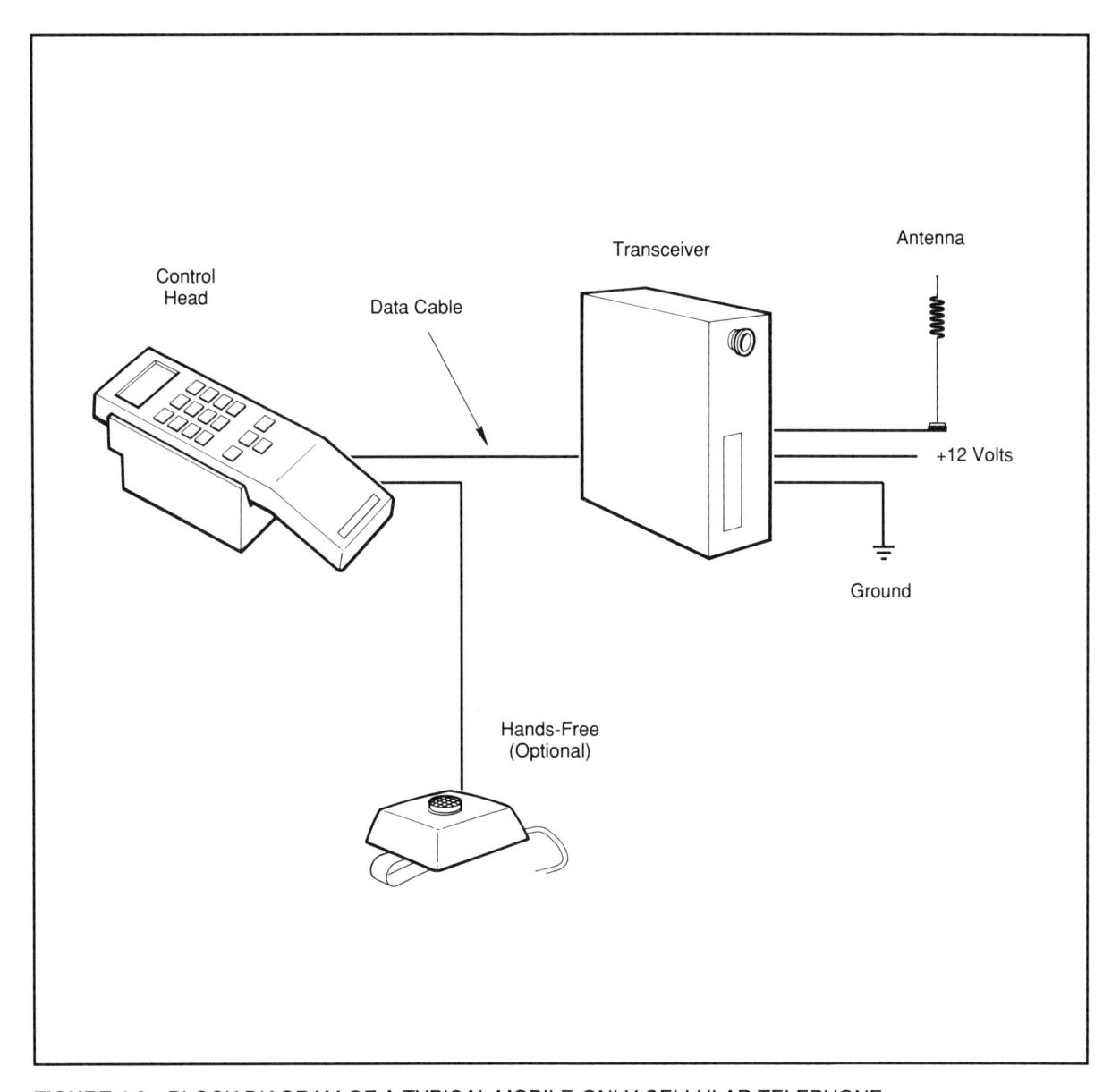

FIGURE 4.2 BLOCK DIAGRAM OF A TYPICAL MOBILE-ONLY CELLULAR TELEPHONE

3. **Data cable.** This connects the control head to the transceiver. The cable allows the two pieces to communicate and thus for the conversation to take place.

4. **Power cable.** This provides power to the phone from the car's battery.

Transportable and hand-held portable cellular telephones (see figures 4.3 and 4.4) operate in the same manner as mobile-only phones with the exception that they contain their own power source. Transportable phones are larger than hand-held portable phones and usually have output power levels of 3 watts. Hand-held portables operate at 0.6 watts.

FIGURE 4.3 TRANSPORTABLE CELLULAR TELEPHONES

Transportable Cellular Telephones

Transportable cellular telephones were developed to fill a gap left between mobile-only units and hand-held portable units. Users wanted the ability to carry their cellular telephones outside of their vehicles. Hand-held portable units were designed to satisfy that need; but, unfortunately, most cellular systems (during the early years of cellular) did not have cells sufficiently close together to support portable use. As a result, transportable cellular phones were developed. They have a full 3 watts of power output and their own battery systems. The early ones were nothing more

FIGURE 4.4 HAND-HELD PORTABLE CELLULAR TELEPHONES

than mobile-only units mounted in briefcases with batteries. The term "transportable" was rather laughable, since some of the earlier units weighed over 25 pounds. Not many people were willing to carry that much weight around. I remember an advertising campaign we considered using (jokingly, of course) for these briefcase units. Our slogan was "Save $50 per month with our new briefcase cellular telephone." Users could save the $50 by buying the briefcase telephone and quitting their health club (since they would get a good workout just by carrying the phone around).

Transportable cellular phones still thrive in today's market. In new systems, or for people who need cellular operations in fringe areas, they represent the only solution for a user who wants to take the phone with them.

A transportable phone is basically a very small cellular transceiver, handset, and battery with a handle. These phones range from the very simple to complex. Some units are nothing more than a mobile cellular phone in a canvas bag. Other manufacturers have produced very sophisticated units with power sources in attractive packages. The same features that are important for a mobile-only phone are also important for a transportable cellular phone. Here are some additional important features to look for in a transportable cellular phone.

Operating Time is the single most important feature of a transportable phone. There are two operating times to consider: stand-by and talk. The stand-by time is the time the phone will operate while waiting to receive a call. When the transportable is waiting to receive a call, it is said to be in stand-by mode. This means only the receiver and computer logic are operating. Both of these systems consume relatively small amounts of power. When the user places or receives a call, the transmitter is turned on. The transmitter consumes a great deal more power than the receiver and computer logic. The transceiver is on only while a call is in progress. Talk time is the time the unit will operate while in transmit mode. The reason these two specifications must be mentioned is that the phone will consume more power in transmit mode than in stand-by mode.

Current Consumption directly affects the operating time and weight of the phone. The more current the phone consumes, the larger the battery must be for the same operating time. The larger battery makes for a heavier transportable telephone. When selecting a transportable telephone, be sure to look for equipment with low transmit and stand-by current consumptions.

Weight is an important feature for the person who has to carry the telephone. Ask anyone who carried around one of the earlier 25-pound "briefcase" telephones. There are two factors that affect the weight of a transportable. The first is the level of technology of the transceiver. If the manufacturer is using the latest LSI (large-scale integration) and surface-mount techniques, their transceiver will be smaller and lighter. It will also probably be more expensive (you don't get something for nothing). The smaller, more advanced transceiver will presumably consume less current during transmit and stand-by than a larger, less advanced unit. When comparing weights of different transportable phones, make sure that you are comparing similar units. It is, of course, easier to manufacture a lighter unit with less operating time.

Power Output is an important feature to consider because the more power the transportable is capable of delivering to the antenna, the more power it will consume from the battery. One of the methods that some manufacturers have used to increase the operating time is to reduce the power output of the transmitter. This will also lower the manufacturing and retail price of the equipment. Some transportable units will operate only at a lower power level, around 1.2 watts. While this lower power output may be permissible under some circumstances, it is likely that the lower power will cause problems. A good compromise is to purchase a unit that has user-selectable power output. This allows the user to lower the power when the situation will permit it. Of course, this flexibility costs money. A unit that offers the user the capability of manually adjusting the transmitter power output will likely be more expensive.

Power Sources are the means by which the transmitter gets its power. Transportable cellular phones can be powered in a number of ways. Listed below are a few of these ways.

1. The transportable unit gets its power from the internal battery when it is not connected to an external source. These batteries are available in many varieties. Lead-acid and gel cell batteries are adaptations of the batteries that are used to start your car. They offer a wide range of operating temperatures and no battery memory problems. These types of batteries are very difficult to overcharge. This permits them to be left on chargers for long periods of time without damaging them. These lead-acid and gel cell batteries are the most popular with the equipment manufacturers. Some manufacturers use removable nickel-cadmium cells. While they do not offer the operating temperature range of gel cells, they can easily be replaced by users (so they can carry several with them) and are fairly inexpensive. Some very inexpen-

sive transportable cellular telephones are not sold with batteries. They obtain their power from external sources. By not having an internal battery, they are a great deal less expensive and lighter than their counterparts.

2. AC charger and power supplies permit transportable units to have their internal batteries recharged and to be operated from AC power sources. These chargers convert 120 volts AC (or house current to those of us who speak "English") to 12 volts DC, which the transportable phone can use. It is a real advantage if the charger not only charges the phone's internal battery, but also operates the phone at the same time. This saves customers the trouble of removing the battery when it needs to be charged and also still allows them to operate the phone at the same time.

3. Since transportable telephones are frequently moved from car to car, it is convenient if there is a quick way to use a vehicle's 12-volt DC power supply to operate the phone. The best way to do this is to use a cigarette lighter adapter. This cable plugs into the cigarette lighter in a car and directs the power to the cellular unit. Some phones even have special circuits that allow the internal battery to be charged from the vehicle's 12-volt DC power supply. This is done by either stepping the voltage up to higher potential or charging the internal battery in two steps. Some transportable phones have elaborate state-of-charge indicators to inform the user of how much power remains in the battery. It is very difficult to get an accurate state-of-charge indication from either nickel-cadmium or lead-acid batteries. A service manager should make sure that the customer knows that these indicators give only a rough estimation of the amount of power left in the battery. Most transportable phones do, however, have an audio alert that lets the user know when the batteries are almost out of power. Usually, these indicators are fairly accurate (and consistent).

Mobile Installation Kits permit the transportable phone to be mounted in a vehicle like a mobile-only unit. While this may sound like a contradiction to the design of a transportable, it is not. There is a natural tendency to place a transportable unit on the seat of a car and operate it. Most manufacturers of transportable equipment recommend that it at least be used with a magnetic-mount roof-top antenna, but in reality this is rarely done. The solution is a mobile installation kit. To be successful, the kit should be simple and inexpensive. A customer will not want to spend a great deal to mount the phone in the car after spending the extra money for a transportable unit. The kit should also be easy for the customer to use. A properly designed kit should make it very easy to install and remove the phone from the car. These kits are also useful for customers who want to install their phones in several different cars.

Carry Bags are great for protecting the transportable phone from outside environmental elements. A carry bag is handy in situations where the customer doesn't want to advertise the presence of a cellular phone to the world.

Auto Shutoff is offered by some manufacturers as a power-saving feature. The auto shutoff can be programmed by the user (or installer) to shut the phone's power off after a certain period of time. The purpose is to save the unit's battery power if it is accidentally left on. This feature can be disabled.

Hand-Held Portable Cellular Telephones

Hand-held portable cellular telephones are very similar to transportables in concept. There are two principle differences, however. The most obvious difference is size. Hand-held portables are complete, stand-alone cellular telephones that fit in the user's hand. They look similar to the cordless telephones that are sold for use at home.

The second difference is power output. Hand-held portable phones are designed to operate at 0.6 watts output. The output is restricted for power management and health reasons. It would be impossible for a battery that was small enough to be practical in a hand-held portable to supply the power required to operate a transmitter at 3 watts. The state of battery technology does not yet support such a requirement.

The features that are important in a transportable unit are also important in a hand-held portable unit. Installation kits are about the only area where the two differ. Two types of hand-held portable vehicle installation kits are currently offered today.

Nonamplifying Kits improve the performance of a hand-held portable by providing the user with an external antenna. By placing the transmitter's antenna outside of the vehicle, the performance of the phone is significantly improved.

64 CELLULAR TELEPHONE EQUIPMENT

As shown in figure 4.5, the hand-held portable is used as a control head and transmitter. The vehicle furnishes power from its electrical system. The cost for a kit of this type is relatively low. This simple kit will offer a great deal of improvement in performance over a hand-held portable being operated from inside of a car. By placing the antenna on the outside of a car, the phone will benefit from improved antenna performance and at least 3 db of gain from the antenna. Some manufacturers offer 5-db antennas with their installation kits to get a little bit of extra gain. Some of these installation kits are sold with separate control heads and hands-free options. In these cases, the hand-held portable is used only as a transmitter and is usually mounted in the vehicle's trunk. The control head provided with the installation kit is much lighter than the hand-held portable and is easier for the customer to use.

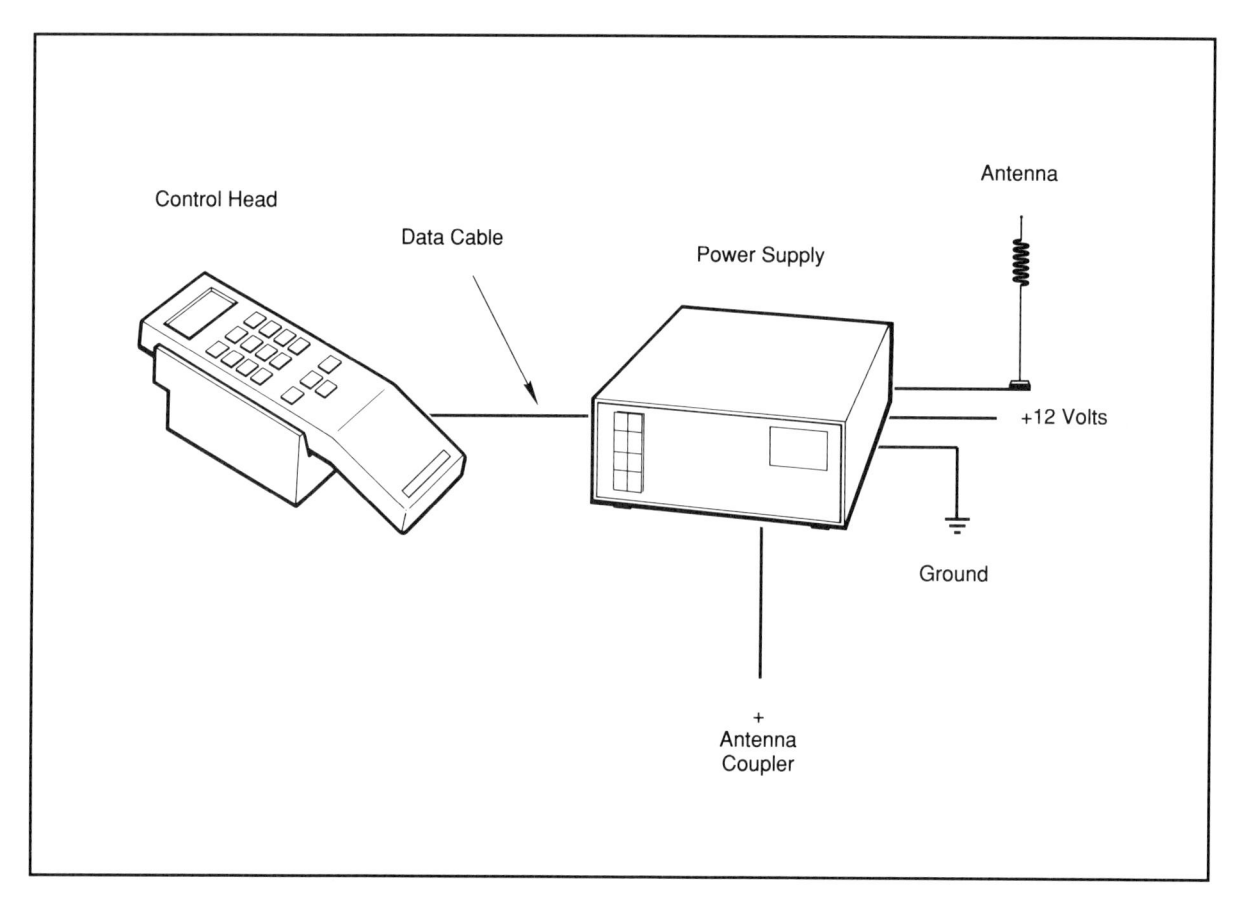

FIGURE 4.5 NONAMPLIFYING INSTALLATION KIT FOR A HAND-HELD PORTABLE TELEPHONE
A nonamplifying installation kit offers the advantage of having an external antenna at low cost.

Amplifying Kits bring the power output of the hand-held portable up to a full 3 watts. In figure 4.6, we see an external power amplifier, which brings the power of the unit up to 3 watts. The hand-held portable is mounted inside the vehicle cabin, and its keypad is used to control the functions of the phone. A "dummy" handset is used instead of a full-featured one. In figure 4.7, we see a variation on this theme. The hand-held portable is mounted in the trunk of the vehicle, along with a power amplifier. A full-featured control head is mounted in the vehicle cabin. This installation looks nice and allows quick removal of the hand-held portable for portable use. Unfortunately, it is very expensive. The

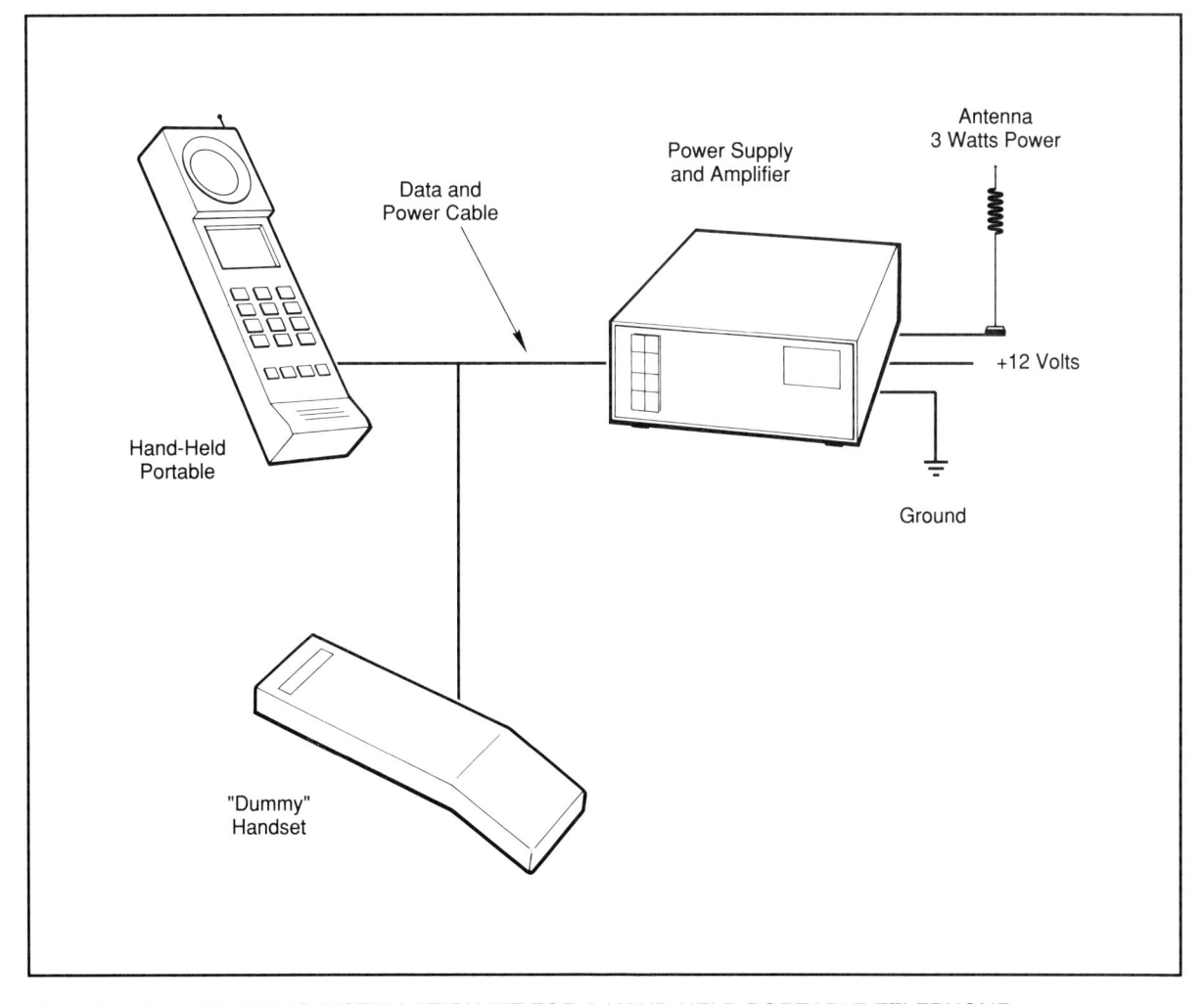

FIGURE 4.6 AMPLIFYING INSTALLATION KIT FOR A HAND-HELD PORTABLE TELEPHONE
An amplifier provides a full 3 watts of power output and supplies power to the hand-held portable phone.

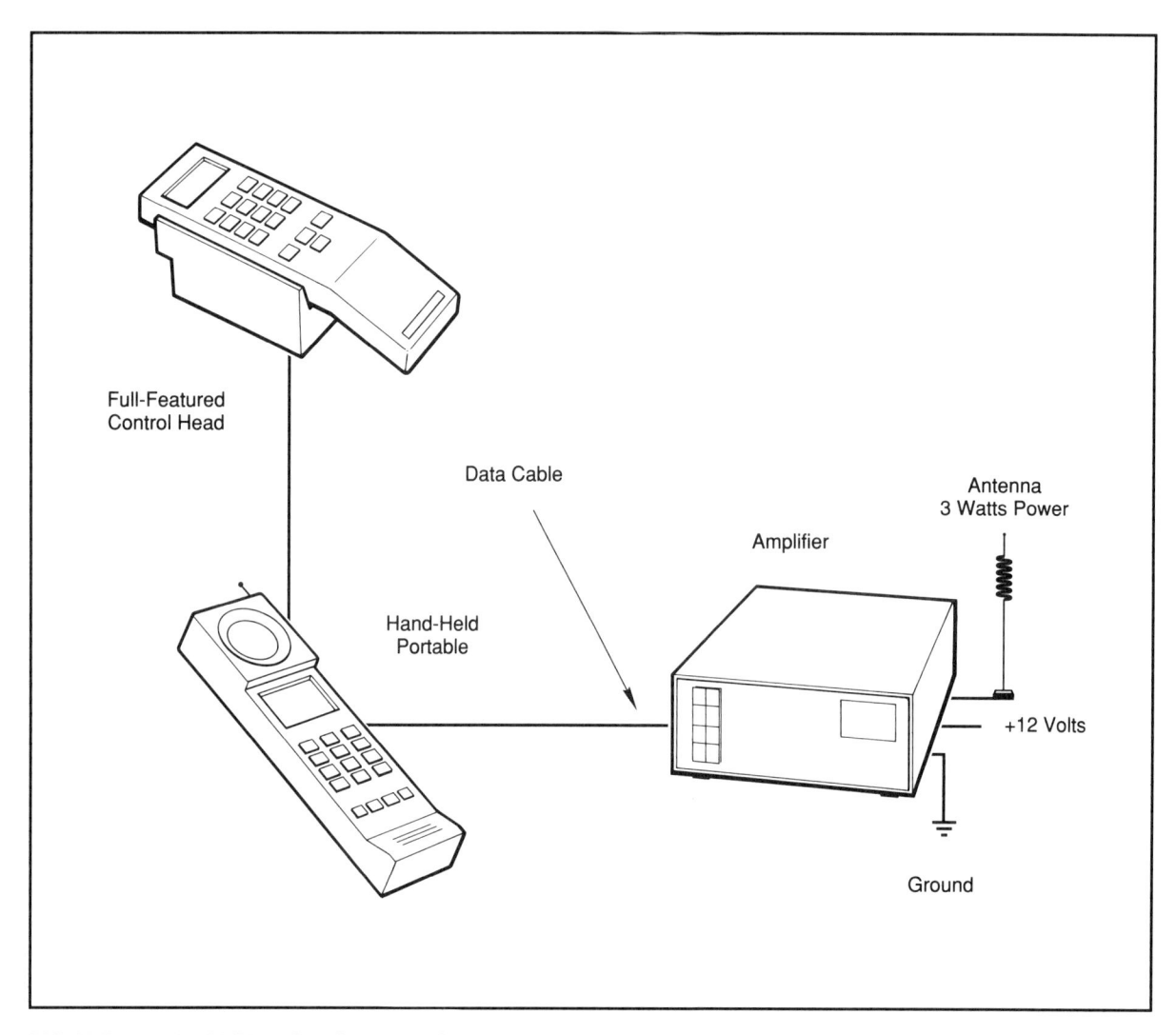

FIGURE 4.7 AMPLIFYING INSTALLATION KIT FOR A HAND-HELD PORTABLE TELEPHONE
The hand-held portable phone is located in the trunk, and a control head is used to operate the phone.

power amplifier must contain a duplexer and costly power transistors.

Another variation is shown in figure 4.8. Instead of using an expensive power amplifier, the hand-held portable has its output increased when it is used with the vehicle kit. The power transistors are capable of operating at an output of higher than 0.6 watts, although they will consume a great deal of power while doing so. In a portable environment, power consumption is important. In a vehicle kit, the power is supplied by the car's electrical system;

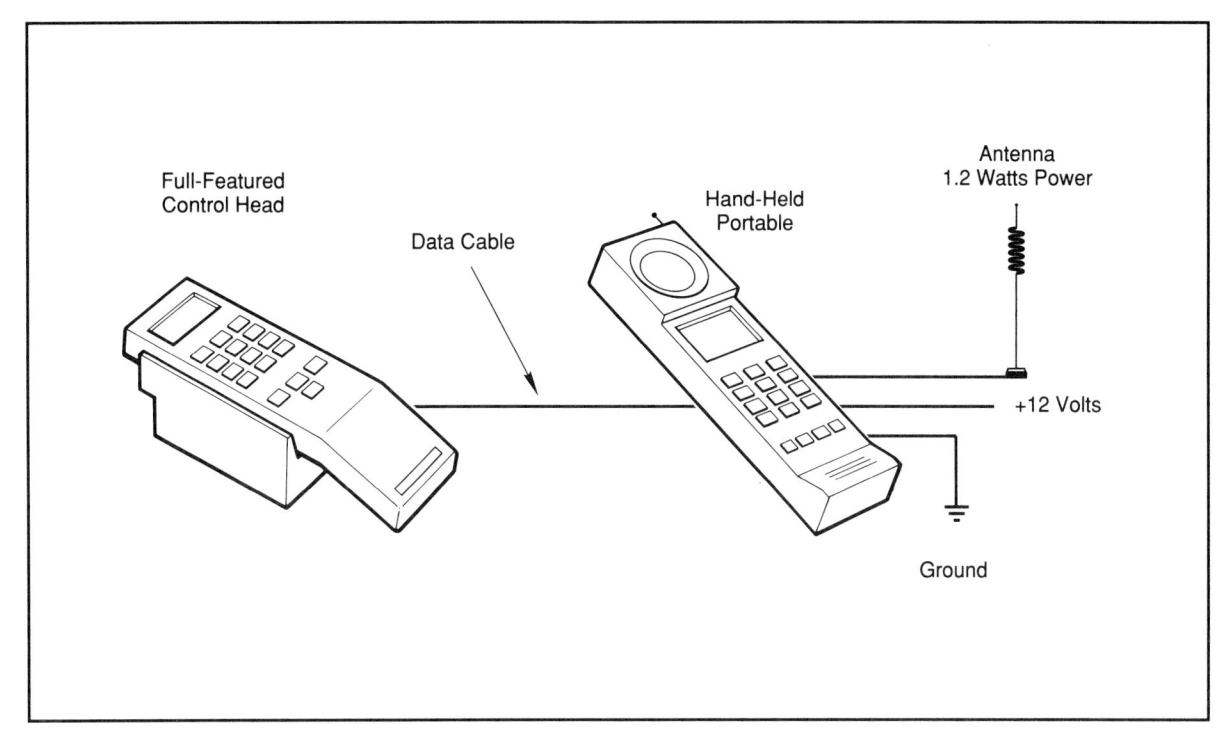

FIGURE 4.8 VEHICLE INSTALLATION KIT FOR A HAND-HELD PORTABLE TELEPHONE
The power of the hand-held portable phone is "turned up" to 1.2 watts.

therefore, power consumption is no longer a consideration. By using this approach, a considerable amount of expense is spared.

Cellular Telephone Features

Cellular telephones are sold with a wide variety of features. Some phones offer only basic features, while other more expensive models are sold with everything from "soup to nuts." Here is a list of the more common cellular telephone features.

Speed Dial Memory is one of the more important features in a cellular telephone. The speed dial memory allows the user to store a telephone number in the phone. Instead of dialing a frequently called number, such as 1-212-555-1234, the number can be stored in the phone's memory. Most cellular telephones assign two-digit numbers (such as 01, 15, 35, etc.) to a memory location. A telephone number is stored in that location and can be quickly recalled at a later time. The cellular telephone owner simply presses the recall (RCL) button, followed by the location (i.e., 02), and then the SEND key. Some of the more sophisticated cellular telephones offer 100 locations to store telephone numbers.

System Select Switch gives the customer manual control over which carrier is used when roaming. While this feature is not important when operating in the customer's home area (because the alternate home carrier will not let the customer roam), it becomes very important when the customer wants to use the phone in another city. Each city (at least, most of them) has two service providers. The customer can choose which carrier to obtain service from. The A/B switch, or system select switch, permits the customer to manually choose a carrier. The higher the quality of the cellular telephone, the easier it is for the customer to perform this function. Some units even let the user know which system is being scanned by indicating the carrier type on the display. Some manufacturers have a different approach and choose to do this by flashing the ROAM light when the phone is scanning a system other than a home-type system. A good example is a customer who drives from New York City to Washington, D.C. In New York City, he is a NYNEX Mobile customer, but in Washington, D.C., he decides to use Cellular One service. The customer's ROAM light would flash on and off to indicate that he is using a nonwireline service. If the customer chose to use the Bell Atlantic system, the ROAM light would remain lighted constantly.

Expanded Channels are offered by most cellular telephone manufacturers. When cellular telephone service was first offered, only 666 channels were available. The FCC held spectrum in reserve. In some of the larger markets, the FCC has allocated this additional spectrum, so that a total of 832 channels is available. (Wireline and nonwireline each get half.) Earlier cellular telephones didn't have the ability to use this additional spectrum since it was not available then. Most of the equipment being sold today is capable of using the additional channels. For the customer who purchases one of the phones with expanded channels, there is less of a chance of a call being blocked in a busy system. Many older model phones can be upgraded by the manufacturer to accommodate the new channels.

Backlit Keypad and Display are important, since cellular phones are also used at night. When properly done, the backlighting provides a warm, even lighting that blends in with the vehicle's own lighting.

Hands-Free is an important safety and convenience feature in a cellular telephone. A hands-free unit is really a speakerphone that is developed for use in a car. Most states actually have laws that require both hands be on the wheel while the car is being driven. (I guess it is illegal to shift gears or change a tape in the cassette deck.) The hands-free feature consists of a microphone that is usually mounted on the sun visor. The call-monitoring speaker in the control head allows the driver to listen to the called party without picking up the handset. The driver's conversation is picked up by the microphone. The installer must remember an important fact about hands-free systems: they all sound very hollow. The hands-free microphone picks up not only the driver's voice but also all of the noise from the vehicle's interior. If this is explained to the customer before using the device, there is usually not a problem. It is important for the customer to have realistic expectations about the operation of the hands-free unit. Some units are offered with an external speaker. While the external speaker requires mounting (which necessitates extra installation time), the phone's performance is usually significantly enhanced.

Alphanumeric Operation is a feature that is usually found on higher priced phones. It allows alpha tags to be associated with a telephone number. Instead of storing Aunt Millie's telephone number in memory location 58 (and trying to remember a month later where it was stored), the number can be stored under "MILLIE." This type of number storage and retrieval makes a great deal of sense. The cellular phone user only has to remember the name of the person to be called and then recall the name from memory. Alphanumeric storage of telephone numbers allows a cellular telephone with 100 memory locations to be useful. There are a number of different schemes used by different manufacturers for alphanumeric storage and retrieval. Some are better than others. It is a good idea to look at several different ideas before settling on one method.

Call Timers allow the user to keep tabs on the cellular telephone bill before it arrives. Most phones offer at least two types of call timers: *last call* and *total*. The last-call timer displays the length of the last call. The total-call timer displays the

70 CELLULAR TELEPHONE EQUIPMENT

cumulative time, much like a car odometer. Some advanced cellular telephones will offer additional call timers.

Limousine Option is an important feature if the installation facility plans to sell to the high end of the cellular buyers (and that's where the money is). A limo kit, which is illustrated in figure 4.9, permits two control heads to be attached to one transceiver. Limo kits can also be used in large boats and rural installations where more than one control head is required.

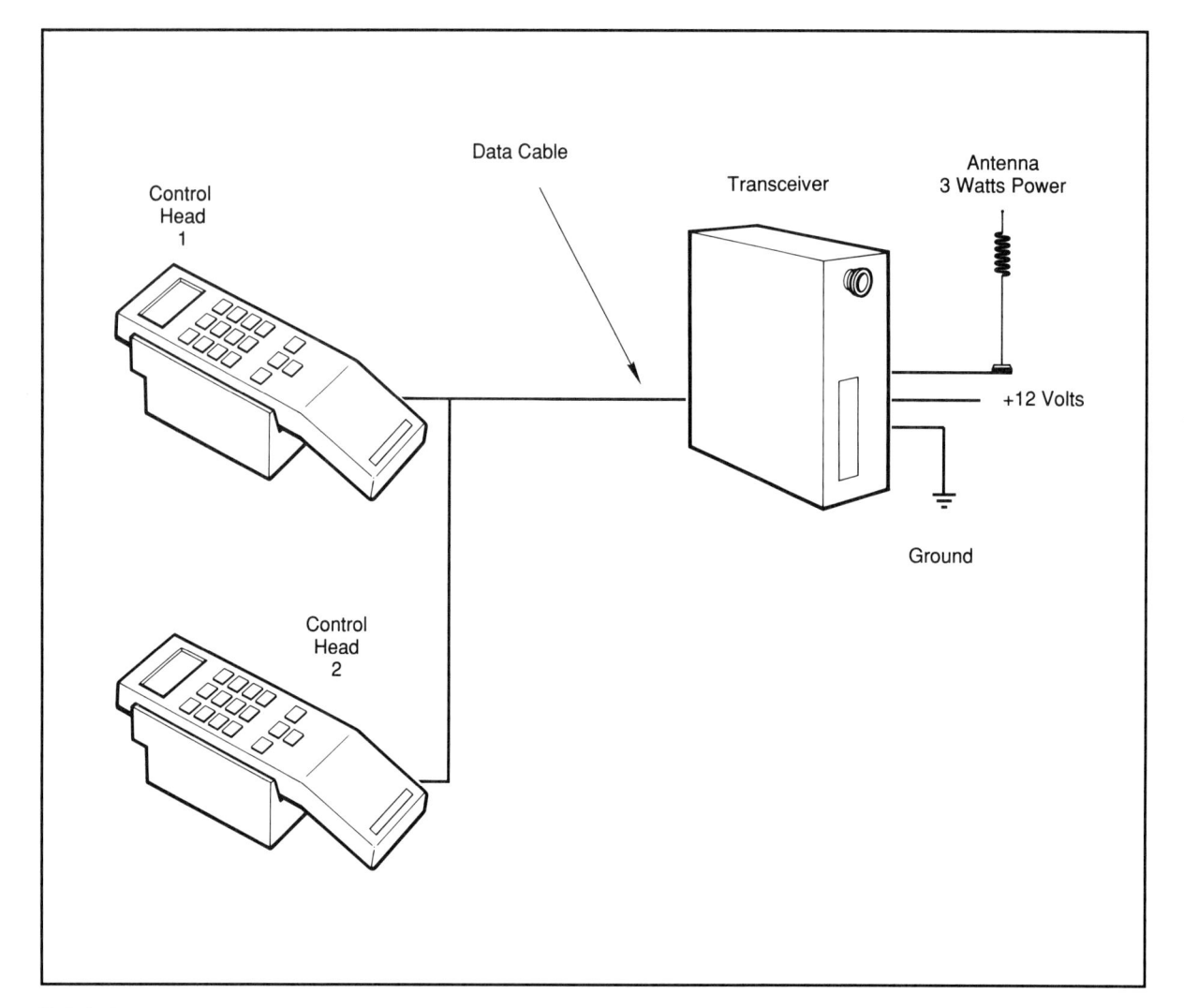

FIGURE 4.9 A LIMO KIT
A limo kit allows the customer to connect two control heads to one transceiver.

Attaching two control heads to one transceiver can be accomplished in a number of ways. Some manufacturers offer junction boxes. This approach is simple but can be expensive, since someone has to pay for the junction box. Another approach is a "T" cable. This is the nicest in my opinion, because it is the least expensive and requires almost no extra work by the installer. Either of the above approaches is acceptable. The only approach I don't consider acceptable is to splice the cable. Any manufacturer who asks an installer to do this does not really offer a limo kit. Splicing the cable is asking for trouble. It is a great way to damage the phone if it is done improperly, and it is very labor-intensive.

Signal-Strength Indicator can be a useful feature on a cellular telephone; it allows the user to determine if there is sufficient signal to place a call. While all cellular telephones have indicators to tell users if they are out of range, some units offer an improvement on this concept. The signal-strength indicator gives the user some kind of idea of the level of signal present. The indicator is usually some kind of addition to the telephone's display. Some manufacturers display a relative number (accessed with a special function command), and others display a number of bars. The more bars that appear, the more signal that is available. The signal-strength indicator can prevent a call from being made (in a marginal area) that has a high probability of being dropped.

Mute Switch momentarily disconnects the microphone from the telephone's audio circuit. This lets the driver of the car say something to a passenger without the person on the other end of the telephone hearing the comment.

Ringing Tone should be variable. Since the level of noise in a car is not constant, the ring level needs to be adjusted to cope with the changes. When the windows are rolled down or the stereo is played loudly, the ring can be turned up. Most phones have this feature.

Lock is one of those features that has been expanded upon and varied almost to infinity. The first cellular telephones had the ability to electronically lock. This prevented someone other than an authorized user from making calls—a rather handy feature when your son or daughter takes the car for an evening. While the original feature was simply lock and unlock, today's cellular telephones offer a wide range of features. Here are a few examples:

1. COMPLETE LOCK. The cellular phone will not make or receive calls. The phone is unlocked by entering a special "unlock" code that is programmed into the phone during installation.
2. RECEIVE ONLY. This allows the user to receive calls but not to place any. This feature is handy in some limousine or dispatch applications, or if you loan your children your car but don't want them to make calls.
3. CALL FROM MEMORY. This has many business applications. The phone can be used to make calls only from the speed dial memory. It will not allow the user to dial any other calls. The phone can receive incoming calls.
4. 911 ONLY. Some cellular phones can be programmed to dial only the operator and 911. The operator will know that the phone trying to make the call is a cellular unit and will not allow long-distance calls to be made without a credit card.

Some phones allow users to change their own electronic lock code. This is done through the use of a security code that is programmed by the dealer. It is a good idea for the dealer to use the same security code for each cellular telephone they program. This is done in case the customer forgets his/her "new" unlock code. I know this sounds as if it is not very secure, but remember that we are locking a telephone, not Fort Knox.

Number Display type is a matter of personal opinion. There are two basic types of displays that are in use. The first, and least popular, are LED (light-emitting diode) displays. They offer a few advantages over their counterpart: easy readability at all angles and good visibility at night. The second type, and most popular, are LCD (liquid crystal display) displays. They offer a few advantages over the LED type. The most significant advantage is good visibility in daylight. LED displays tend to wash out in direct sunlight since they emit light. LCD displays reflect light, so they work very well in direct sunlight. Also, they can be backlighted rather nicely so that they can be viewed in the dark. Unfortunately, they are difficult to view from some angles. The new "supertwist" LCD displays should resolve this problem and make LCD displays the choice of most manufacturers.

As long as we are talking about displays, we might as well mention a few other important items. The first is the number of digits displayed. Many older models of cellular telephones displayed only seven digits. If a long-distance call was dialed, the area code disappeared. This was not very desirable. Almost all the newer models offer ten-digit displays. Even this is not enough to display an international call. Some phones even group the dialed number to make it a bit easier to read. For example,

8135551212

is much easier to read when it is displayed as

813 555 1212

Volume Controls are offered on all cellular telephones. They should be conveniently located so they can be easily used. These controls should control not only the volume in the earpiece of the telephone but also the volume of the call-monitor speaker. Additionally, some manufacturers use them to control ring and hands-free volume.

Operating Temperature is usually an important consideration only to people living in extreme climates. Check the manufacturer's specifications to be sure the phone will work in your particular climate.

Call In-Absence Indicator alerts users that someone attempted to call their cellular phone while they were away from their vehicle. If only a few people have been given the customer's phone number, it can be a good feature. For a salesperson or someone who has quite a few people calling the vehicle, this feature is more practical when used in conjunction with "No-Answer Transfer" (a carrier-provided service that will be explained later in this chapter).

Clear Button clears either one digit or the entire display of a misdialed phone number.

Silent Alert will disable the ringer of a telephone and give a visual display when the phone is being called. While this may sound like a rather ridiculous feature, it has applications in areas where it is not appropriate for an audible ring (such as in a court of law, library, church, security, law enforcement, etc.).

Display Own Number is a very useful feature for purchasers of multiple cellular units. Nothing is more frustrating to customers than to pick up the phone they will be using for the day and not know what the phone number is. The "Display Own Number" feature is usually executed in one of two ways. The first is to assign a memory location (i.e., memory 99 in a unit with 99 memories) to the unit's own phone number. The user then presses "Recall XX" to retrieve the phone number. The second is to use a function key (i.e., FCN #) to retrieve the number of the phone.

DTMF stands for Dual-Tone Multi-Frequency. These are the tones that are sent when the number keys are pressed on a Touch-Tone™ telephone. These tones are used to access services such as voice mail, answering machines, alternate long-distance carriers (Sprint, MCI), etc. A word of caution about the use of cellular telephones with these services: Some of these services (such as home answering machines) will not work with cellular telephones because the cellular phone sends short pulses of DTMF tones. Some services require longer pulses than the cellular phone will provide. Be sure to ask the manufacturer of the cellular phone if their DTMF pulses are long enough to activate all of these services.

ADD-ON ACCESSORIES

There are countless accessories sold by small independent companies that are intended to be used with cellular telephones.

RJ11 Interfaces are offered by a few companies with their cellular telephones. A few independent companies offer interfaces that provide true RJ11 emulation (see figure 4.10). RJ11 interface may sound rather ominous and complicated, but it merely refers to the plug that a POTS telephone plugs into in your home. There are many applications for such an interface. Two of the most important applications are computers and facsimile (fax) machines.

By using the RJ11 interface, a lap-top computer or fax machine can be used in a vehicle. While this may sound a bit ridiculous at first, the salesperson who can "fax" in an order from

FIGURE 4.10 AN RJ11 INTERFACE

the car doesn't think so. And the reporter who has written a story "on location" can connect to his/her car telephone and send the text directly to the editor.

Alarm backups are another good application for an RJ11 interface. Instead of using the landline telephone to report a break-in to the police, the alarm can be interfaced to the cellular telephone. The phone is automatically dialed by the alarm system, and there is no way the alarm can be defeated by cutting the phone lines. Even the power cannot be interrupted if a small backup power supply is installed with the alarm.

These are just a few ideas for applications of RJ11 interfaces. The uses are limited only by the installer's imagination.

Cellular Modems are offered for people who are really serious about data transmission. In figure 4.11, we see a photo of a cellular modem with full error correction. Error correction is important, since the audio path can be interrupted during handoffs or during periods of low signal strength. The modem "holds" the data in a buffer during these periods and waits until a clear path exists before sending the data. In situations where data integrity is important (financial, news, medical, etc.), error correction is a must.

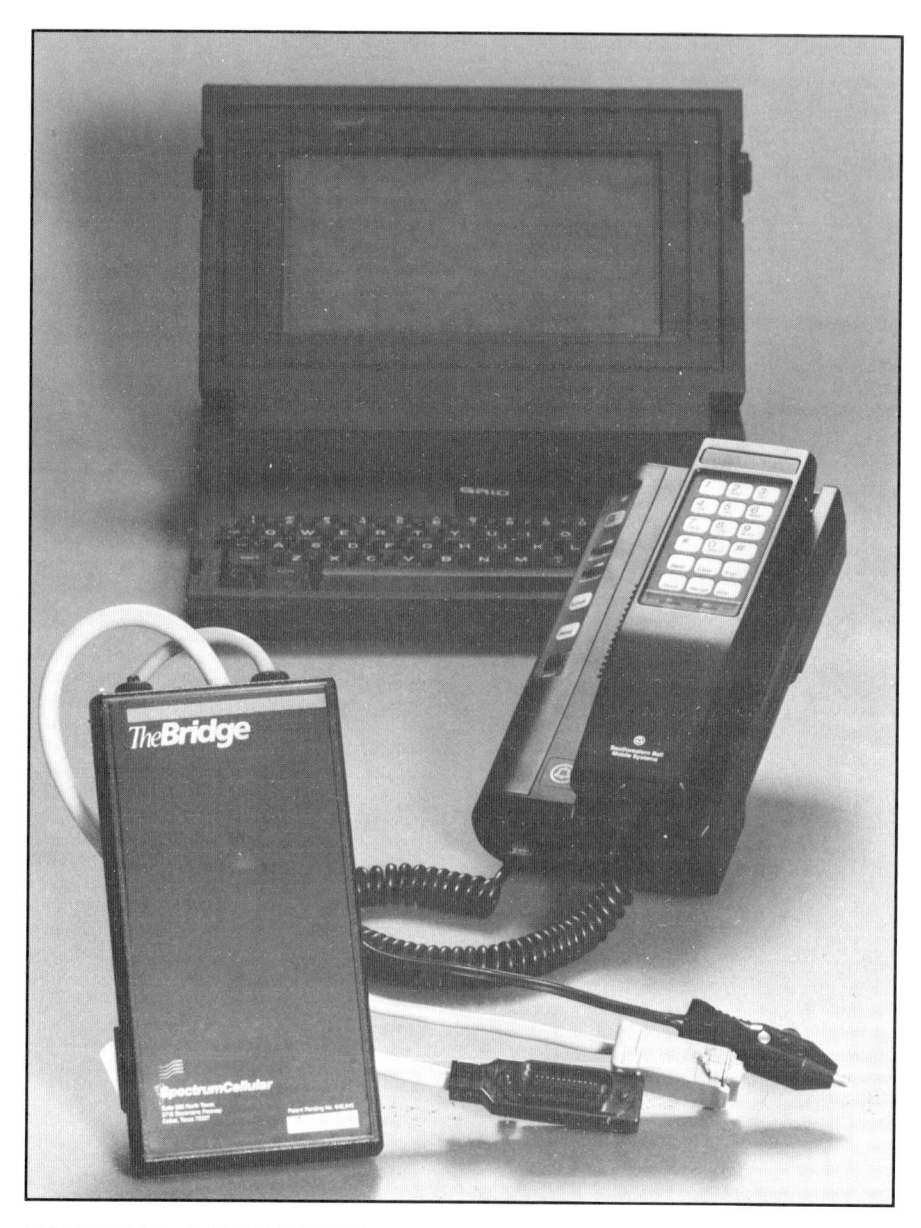

FIGURE 4.11 A CELLULAR MODEM

Voice Recognition is a field where there is still considerable room for development. The science of human voice recognition is still rather young but full of promise. While voice recognition devices are far from 100 percent perfect, they can offer the cellular user some real convenience.

Voice recognition devices are designed to "understand" voice commands from the user of a cellular telephone in order to control the phone. There are two different types of voice recognition devices currently being offered.

1. VOICE-DEPENDENT. This type of voice recognition system requires the device be "trained" to understand one person's voice. The device needs to be "taught" to recognize the way a person pronounces each word in its command set. This needs to be done because people of different ethnic origins or regions pronounce words with unique accents. While the human brain is quite efficient at understanding different accents, machines have problems with this process. The training process familiarizes the machine with the speech patterns of a particular person. The process usually needs to be done only when the voice recognition unit is installed. User-dependent systems are the best-performing systems. They present a few unique problems in that learning how to use them is complicated and the training procedure is complex.

2. VOICE-INDEPENDENT. This type of voice recognition system is designed to understand a particular set of words for all users. This kind of system offers a real advantage in that it is very simple to use and requires no training process. Unfortunately, inexpensive units, such as the ones used with cellular telephone systems, don't work very well. It will probably be a few more years before these units are perfected.

CARRIER SERVICES

Carrier services are special features (or vertical features) offered by the cellular service provider. They are independent of the cellular phone but are activated through the keypad.

1. CALL FORWARDING. This service allows the cellular subscriber to have his/her calls transferred to a particular telephone number. This is a very useful feature for anyone who is in and out of the car a great deal. Keep in mind that the carrier usually charges the subscriber the full air-time charge for any calls that have been forwarded.

2. CALL WAITING. This service alerts the cellular subscriber to an incoming call during a phone conversation. The subscriber hears a tone in the earpiece when there is a call waiting. To accept the call, the subscriber presses the SEND button. The conversation party is put on hold, and the incoming call is answered. The two parties can be "toggled" back and forth by using the SEND key.

3. NO-ANSWER TRANSFER. This service is a variation of the call-forwarding idea. Instead of forwarding all calls to a prespecified telephone number, the calls are transferred only in the event that the phone is not answered. This feature is very useful because it can be activated and left on.

4. VOICE MAIL. This service is used along with call forwarding and no-answer transfer. Voice mail is rather like a central answering machine. Calls from the cellular telephone are sent to the voice mail machine and answered. The caller hears an announcement stating that the cellular customer is not in the car and to please leave a message at the tone. The announcement can be customized by the subscriber. Messages are then retrieved by the cellular subscriber at a convenient time.

CHAPTER 5

Antenna Theory and Selection

ANTENNA THEORY

An antenna is a device that converts electrical energy into electromagnetic waves. Light, X-rays, microwaves, and radio waves are all forms of electromagnetic radiation. These waves move (or, more correctly, propagate) through free space (a vacuum) at about 300 million meters per second. The speed is dependent on the medium in which the waves are traveling, but for our purposes we may assume 300 million meters per second is a good approximation.

Waves don't really move: they propagate. In figure 5.1, we see an analogy that should be familiar to anyone who ever threw a stone into a pond. The stone creates a disturbance in the medium (the water). The disturbance propagates from the center of the disturbance in a circle. The farther out from the center we look at any given time, the smaller the height of the waves. A close look at the water will show the observer that the water is not actually being moved in a horizontal plane. It only moves up and down. As you can see in figure 5.2, the boat only moves up and down: it is not actually pushed anywhere by the wave.

This analogy also holds true for electromagnetic waves. They are only displacements of a particular medium. A more thorough discussion of electromagnetic theory is required to answer some interesting questions (such as how does a wave propagate through a vacuum since there is no medium to displace). Unless you are interested in reading some rather tedious books filled with all sorts of nasty calculus and odd drawings, you will just have to trust me.

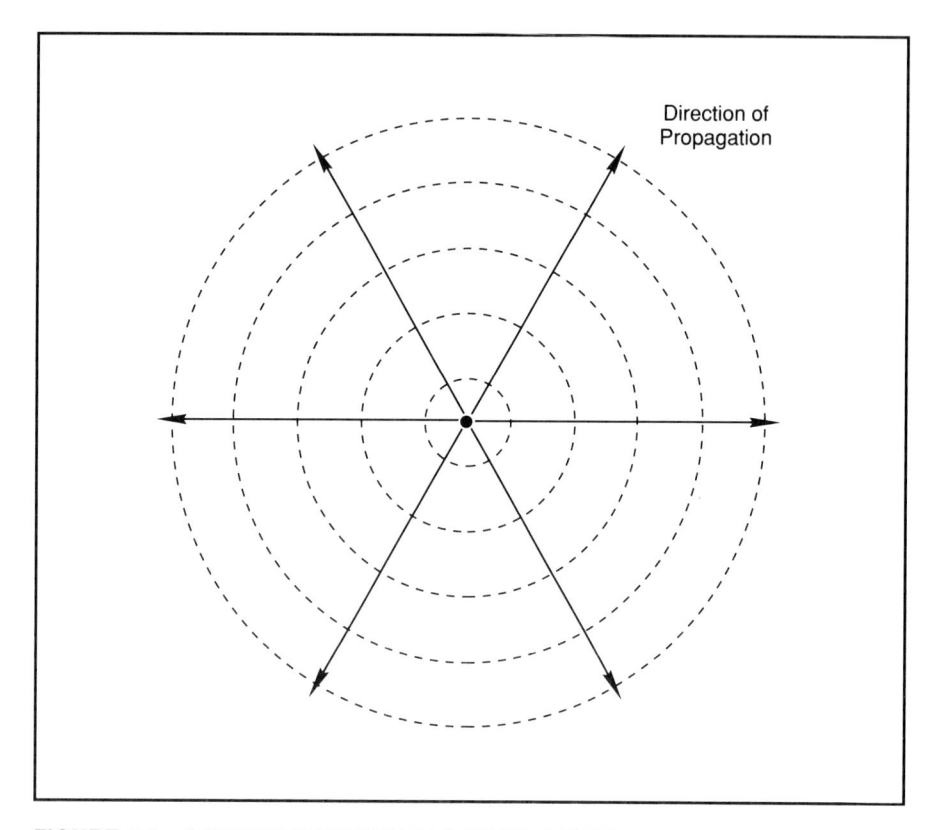

FIGURE 5.1 A STONE DROPPED IN A POND GENERATES WAVES

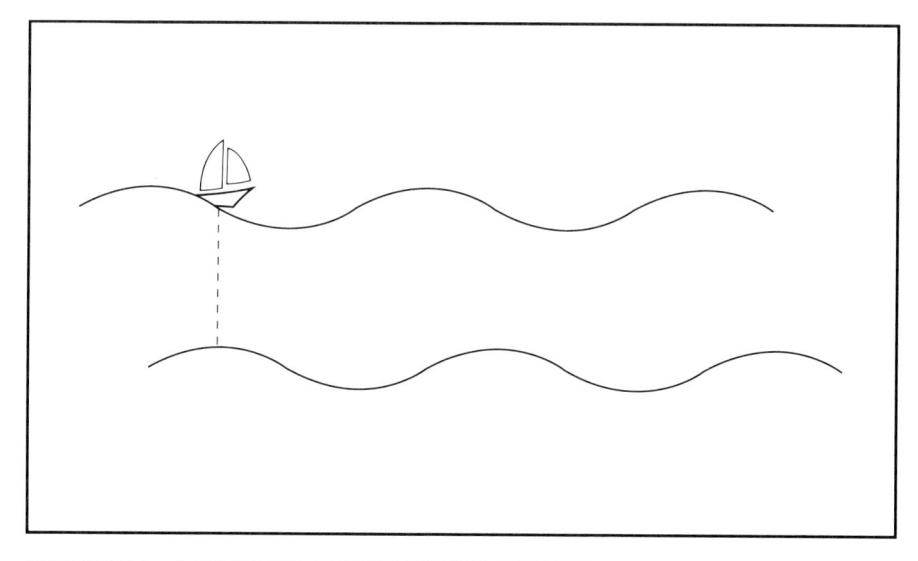

FIGURE 5.2 A BOAT ON A WAVE MOVES ONLY VERTICALLY, NOT HORIZONTALLY

Wavelength

Another very important concept in antenna theory is wavelength. Wavelength is very simply the length of the wave, as the name implies. In figure 5.3, you can see that wavelength is the distance between the first two points of the same amplitude in a wave. To calculate the wavelength of a particular frequency in free space, we use this formula:

$$\text{wavelength (in meters)} = 300 \text{ million/frequency}$$

Since frequency is usually expressed in megacycles (or million cycles per second), we can simplify the formula so it appears

$$\text{wavelength (in meters)} = 300/\text{frequency (in megacycles)}$$

As you can see, lower frequencies have a longer wavelength and higher frequencies have a shorter wavelength. Lower frequencies tend to travel along the surface of the earth, while higher frequencies are more line-of-sight. Cellular telephones operate in the 800 MHz band and are line-of-sight in nature.

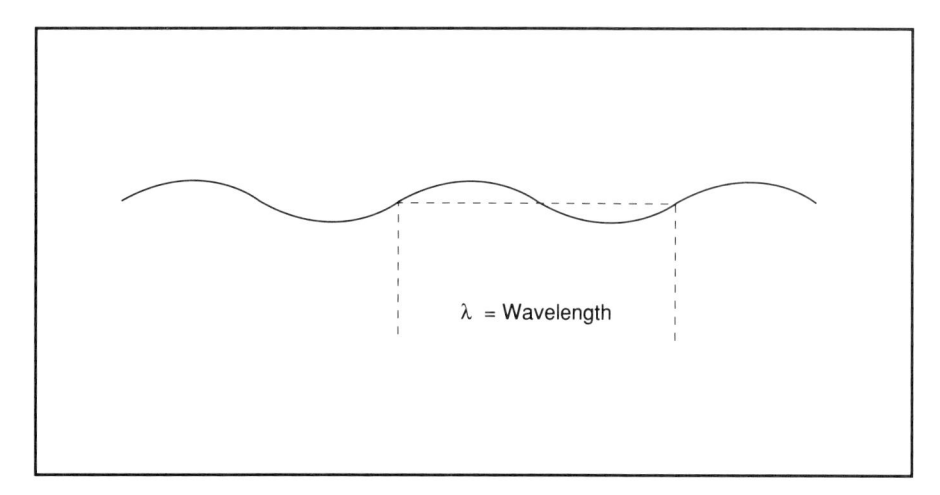

FIGURE 5.3 DISTANCE OF WAVELENGTH
A wavelength is one complete cycle of a wave.

Antenna System

An antenna in any communications system should be designed to most efficiently use the power generated by the transmitter. An antenna system consists of two parts:

1. The transmitter
2. The transmission line

Transmitter. A transmitter is a device used to impress information (in our case, voice or data through a cellular telephone) on a carrier wave and then broadcast that wave through space in the form of electromagnetic information. The carrier wave, which has a distinct frequency, is generated by an internal oscillator and is a simple sine wave (see figure 5.4). The oscillator is usually controlled by a quartz crystal, which is noted for its ability to oscillate at a very exact frequency.

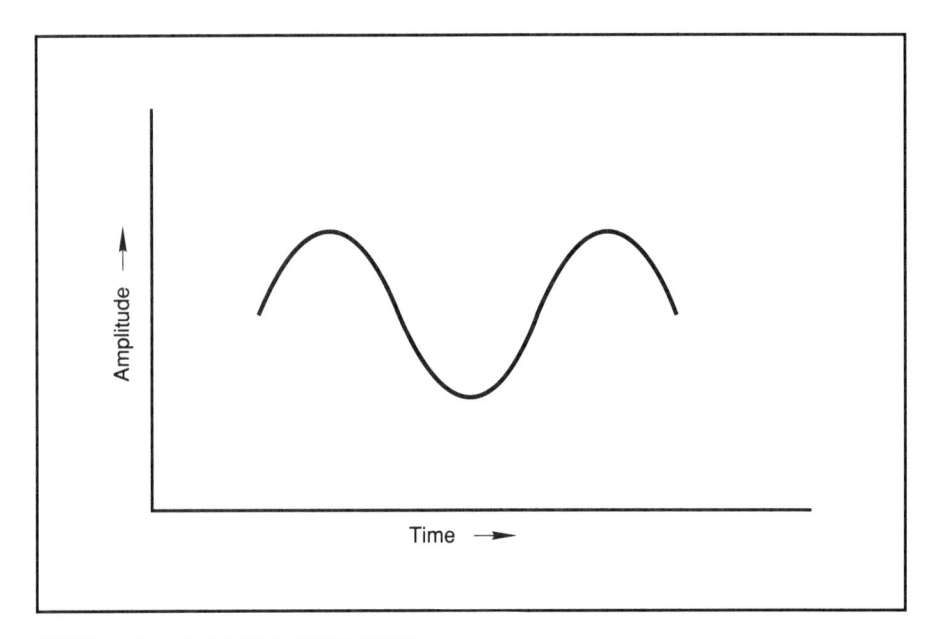

FIGURE 5.4 A SIMPLE SINE WAVE

The information is superimposed on the carrier wave by altering one of the wave's basic characteristics in a process called *modulation*. The most common methods are amplitude modulation and frequency modulation. Cellular telephones operate using frequency modulation. In the case of a cellular phone, the carrier is mixed with the audio signal in a manner so that the frequency of the carrier varies while the amplitude remains constant. After modulation, the carrier signal is amplified to a level of about 3 watts in the case of a cellular telephone. A receiver later picks up the signal and converts it back to audio.

Transmission Line. A transmission line is essentially a pair of wires designed to move electrical energy from one point to another. In our particular case, this transmission line takes the

RF energy and moves it from the transmitter to the antenna so that it can be received by the cell site. This is done with a piece of low-loss coaxial cable. Good coaxial cabling should induce a minimum of electrical loss into the antenna system to ensure maximum power transfer. Since cellular telephones operate at very high frequencies, low-loss cable is essential if the system is going to work properly. RG 58U (this number specifies a type of cable) is the smallest cable that should be used in a cellular installation.

In figure 5.5, we see a schematic representation of a cellular antenna system. In our simplified representation, we see a generator that drives a load through a transmission line. In the case of a cellular telephone, the generator is a transmitter, the transmission line represents the coaxial cable, and the load is analogous to the antenna.

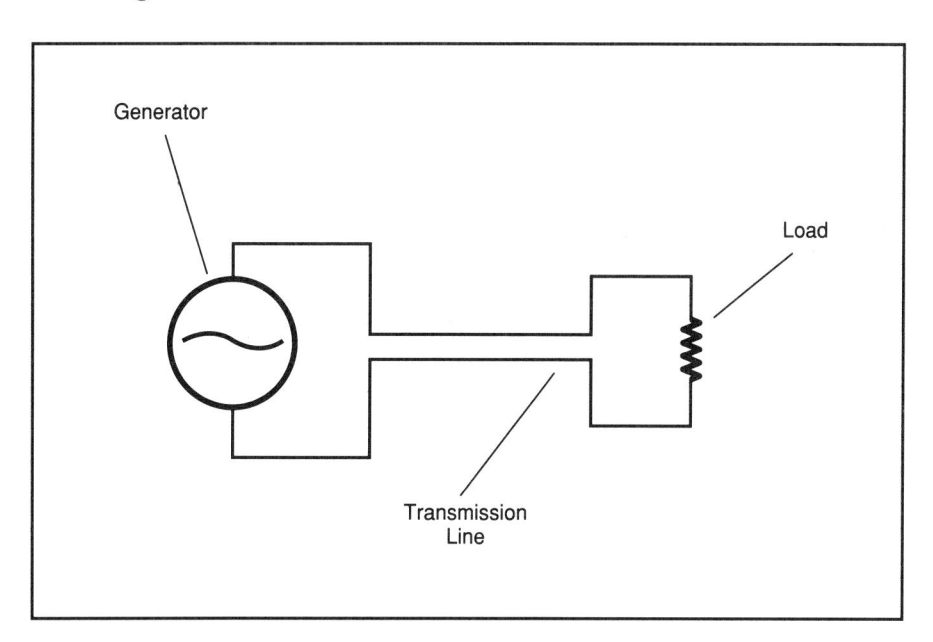

FIGURE 5.5 SCHEMATIC OF A CELLULAR ANTENNA SYSTEM
The generator drives the load through the transmission line.

The transmission line is one of the critical parts of an antenna system because of some if its unique characteristics:

Short circuits in a coaxial cable are disastrous to the system. Consider our transmission line example. The generator sends its electrical energy down the transmission line. At any given time, the two contacts of the generator are opposite polarities. Since our load is a short circuit, maximum current will flow through the load. But since the load has no resistance, there can be no voltage drop across it. Since there is no voltage drop, no energy can be absorbed or radiated.

This can be proved by applying Ohm's law (voltage = current × resistance). If the resistance is 0, the voltage must be 0, regardless of the current. All of the energy is reflected back to the generator. Since the generator sends all of its energy to the load (which is a short circuit), the result is a standing wave on the transmission line. The "bottom line" is that none of the signal makes it to the antenna and the phone doesn't work.

This standing wave can actually be measured by any number of methods. One simple technique (in high-power applications) for measuring a standing wave on a transmission line is to place a neon test light next to the cable. The light will glow at points where the wave is at a peak and will extinguish where the wave is at a minimum. This is also a simple method to measure wavelength, since the light will glow and extinguish once per wavelength.

In the case of a cellular telephone, this short circuit in the coaxial cable prevents any of the power generated by the output section of the transmitter from reaching the antenna. All of the power is, in fact, reflected back to the transmitter. The best-case result is a phone that won't place a call; however, some cellular phones will not tolerate a short circuit, and the output section will be damaged.

Open circuits in a coaxial cable are almost as bad as short circuits. Instead of 0 resistance, we have infinite resistance. The result is that no current flows and the voltage potential across the open load is at a maximum. Again, no power is absorbed or radiated through the load. Instead, all power is reflected back to the transmitter in the form of a standing wave. In the case of a cellular telephone, no current ever reaches the antenna and, as a result, no signal is transmitted. While this situation will usually not damage a transmitter, no power reaches the antenna, and the phone will not place a call.

These problems usually result from bad connector crimping. Proper crimping techniques will be discussed later in this book. These two examples should illustrate the necessity of properly running and attaching connectors to coaxial cable.

Antennas

As stated earlier, the antenna is one of the most important parts of a cellular installation. Its job is to convert the electrical energy generated by the transmitter into electromagnetic energy in the most efficient manner possible. Most cellular antennas are a variation of the Marconi-type antenna. A Marconi antenna is a resonant antenna (see figure 5.6). It is referred to as resonant because its physical length is associated with the frequency it radiates. A quarter-wave cellular antenna is usually a Marconi-type antenna and has a length one-quarter of a wavelength of the frequency of the signal it is transmitting.

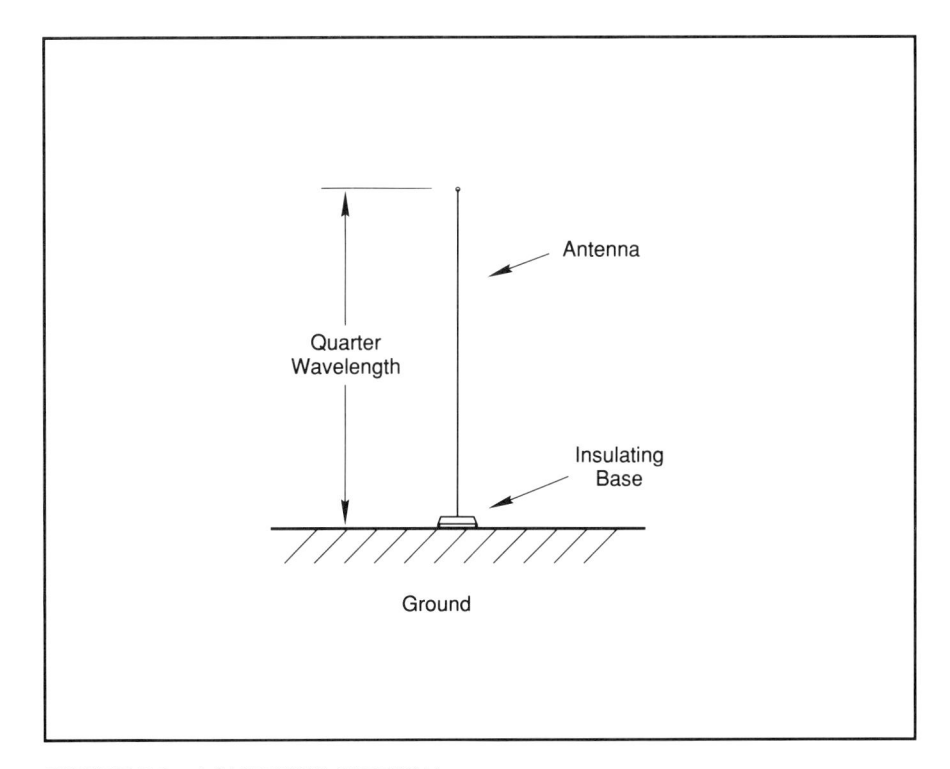

FIGURE 5.6 A MARCONI ANTENNA

The Marconi antenna is usually one-quarter wavelength long. It uses the ground to act as a mirror quarter-wave segment to the antenna and acts itself like a half wave. This is very desirable in a cellular environment, since it keeps the antenna rather short.

In a real world situation, the roof of the car acts as a ground plane when an antenna is mounted there. The antenna is mounted on an insulating base, and the coaxial cable is fed through a small mounting hole made in the roof. By isolating the antenna from the ground plane, matching and consistent radiation patterns are less of a problem.

IMTS mobile antenna installations usually required that antennas be tuned or cut to their specific operating frequencies. One of the facts that must be taken into consideration when designing an antenna is that radio waves travel more slowly in air or in coaxial cable than in free space. This means that the actual wavelength is slightly shorter than theoretical calculations will predict.

A phenomenon known as *end effect* also tends to reduce the actual resonant length of an antenna. This happens because air acts as a dielectric and capacitively couples the two ends of the antenna together. This also reduces the electrical wavelength of the antenna. The antenna is therefore shorter than would be predicted by theoretical calculations. Fortunately, installers of cellular antennas do not have to worry about tuning antennas. The types of antennas available for cellular installations are designed to operate over a wide range of frequencies and are pretuned by the factory.

Although an installer of cellular telephones doesn't need to tune antennas, it is a good idea to understand these relationships and how antennas work. About half of the problems you will experience with cellular telephones will somehow relate to the antenna. The more you understand, the easier it will be to fix the problems.

Antenna Radiation Patterns

A theoretical antenna is a single point in space that occupies no space and has no connector or cable attached to it. It is completely free of interactions with any external objects and radiates equal strengths of signal in all directions (see figure 5.7). In reality, antennas are physical objects with specific dimensions that interact with other objects. As a result, real antennas radiate electromagnetic energy at different intensities in different directions. A Marconi-type antenna mounted on the roof of a car will radiate in a doughnut-shaped pattern (see figure 5.8). This type of pattern is very applicable to cellular communications, since cell sites are usually positioned on the same plane as (or very close to) the vehicle. The Marconi-type antenna also offers the advantage of being simple (and inexpensive) to manufacture and install.

ANTENNA THEORY AND SELECTION 87

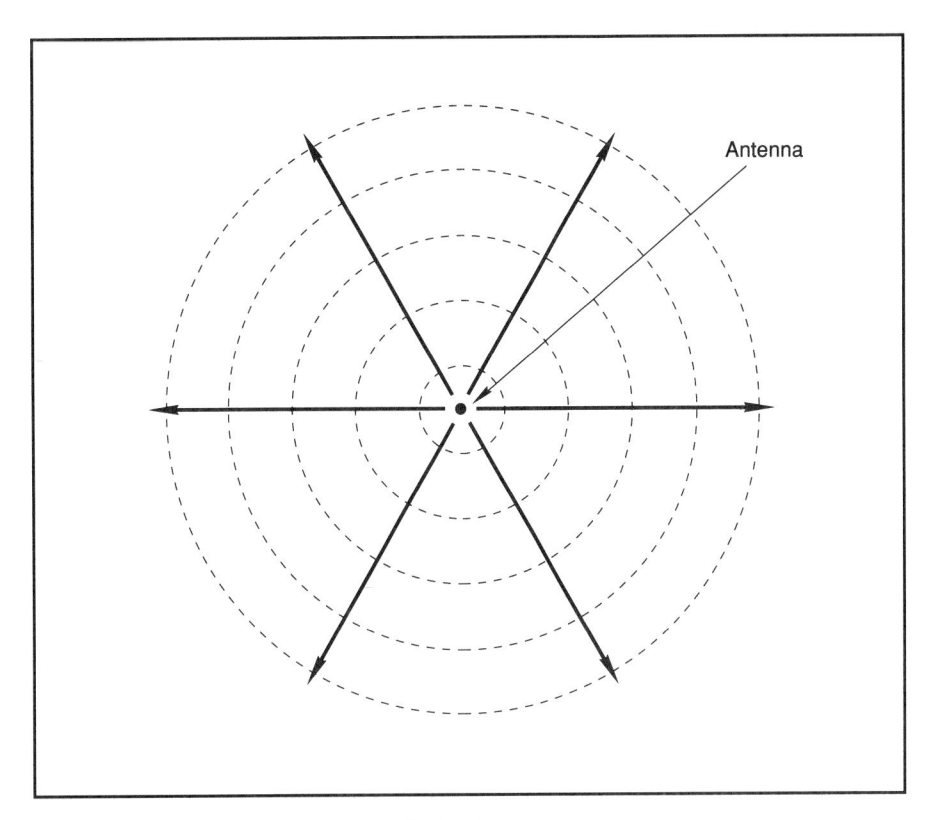

FIGURE 5.7 A THEORETICAL ANTENNA
A theoretical 0-db gain antenna radiates equally in all directions.

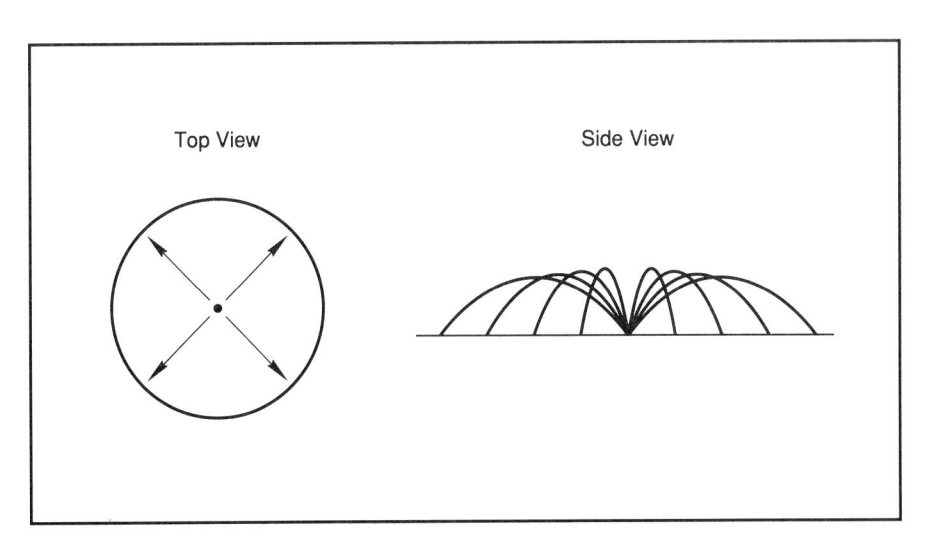

FIGURE 5.8 RADIATION PATTERN OF A ROOF-MOUNT MARCONI-TYPE ANTENNA

ANTENNA SELECTION

Three different types of cellular antennas are used in vehicle installations:

1. Glass-mount (or through glass)
2. Roof-top
3. Elevated-feed

Just about every antenna sold today is a variation of one of these three basic types.

Glass-Mount Antennas

The glass-mount antenna is by far the most popular cellular antenna (see figure 5.9). It offers several advantages over roof-top and elevated-feed designs. The most significant advantage is ease of installation. A glass-mount antenna, as the name implies, is mounted directly to one of the windows of the vehicle. The usual choice is the rear window. A length of coaxial cable connects the transmitter to a coupling box that is mounted to the rear window with a silicone adhesive (similar to clear silicone bathtub adhesive). The coupling box uses the glass as a capacitor to couple the energy to the antenna, which is mounted to the outside of the window.

Theory of Operation. There are two basic glass-mount systems in use today. In figure 5.10, we see the Avante-style glass-mount antenna. In this approach, a collinear radiator was designed to be mechanically substantial enough to withstand the rugged environment of a vehicle antenna. It also needed to maintain an acceptable current/phase relationship across the cellular band. Also, a broadbased coupling circuit was designed to efficiently couple the RF signal through the vehicle's window. The coupler capacitively couples the internal box to the antenna and voltage feeds the base of the radiator. This box is attached to the inside of the window with an adhesive and couples the RF signal to the antenna mounted on the outside of the window. The use of coupling plates in series with the base of the antenna introduces a capacitive effect that shortens the electrical length of the antenna element. This allows the antenna to be cut shorter, a cosmetic benefit. These coupling plates are operated at very high impedances (between 25K ohm and 100K ohm). The internal

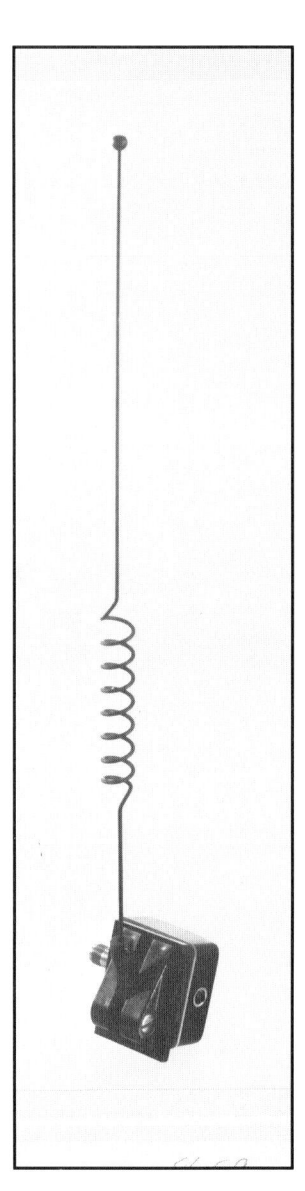

FIGURE 5.9
A GLASS-MOUNT ANTENNA

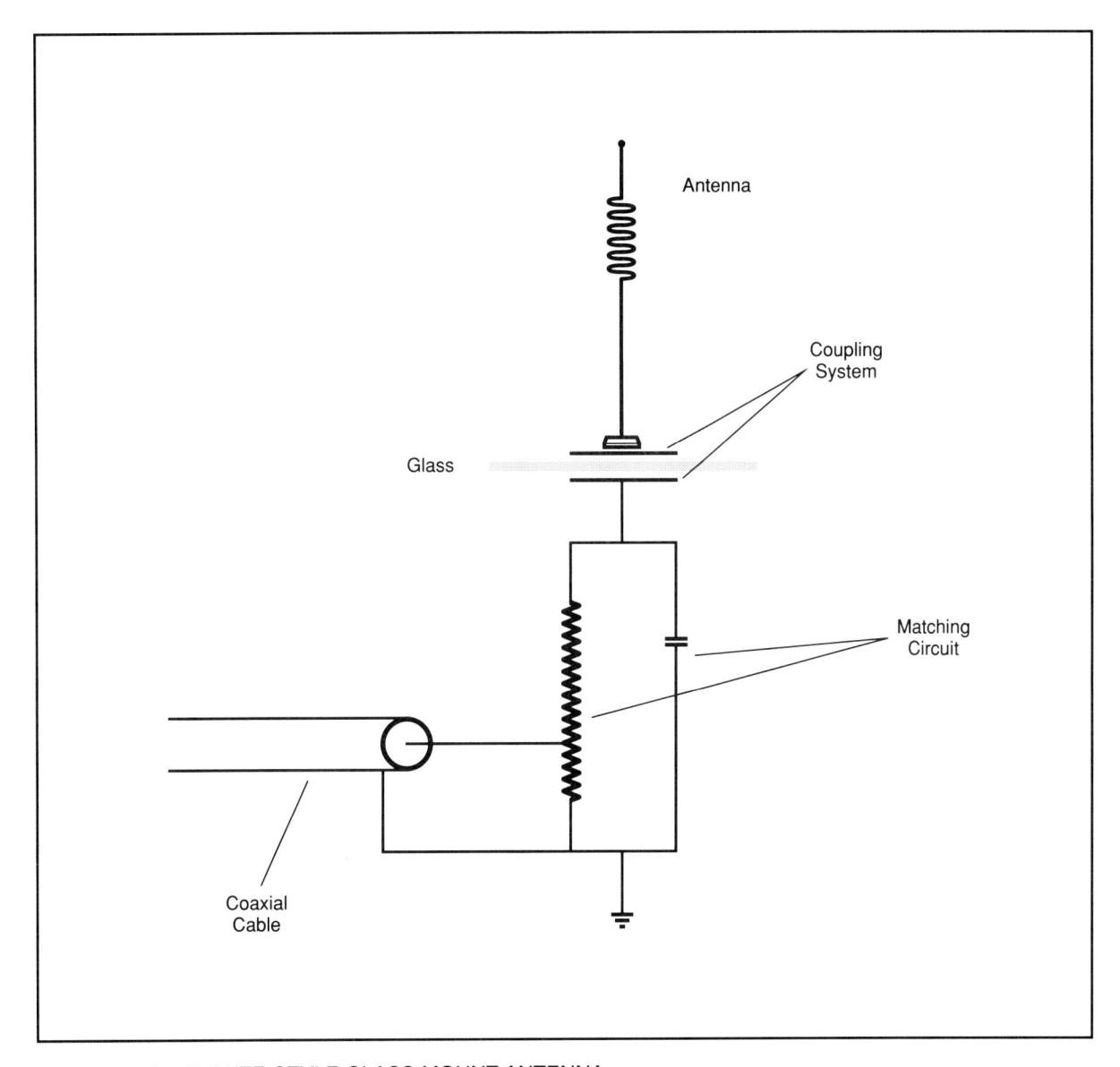

FIGURE 5.10 AVANTE-STYLE GLASS-MOUNT ANTENNA
This is a glass-mount antenna with the matching circuit *inside* the window.

plate is integrated into the coupling unit. The outside plate is the inside surface of the antenna assembly. It provides physical support in addition to coupling the antenna to the transceiver. The 800-MHz signals require a very small coupling surface area to facilitate the transfer of energy to the antenna. The signal loss from this transfer is very small if the antenna is tuned properly.

90 ANTENNA THEORY AND SELECTION

The only drawback of this particular antenna system is a side effect of the high-impedance coupler design. The high impedance causes this type of antenna to be subject to some variation in impedance and radiation characteristics when mounted improperly. However, if care is taken to follow the manufacturer's instructions during mounting, a properly operating antenna will result.

In figure 5.11, we see a variation on this theme. This style glass-mount design feeds the RF energy directly through the

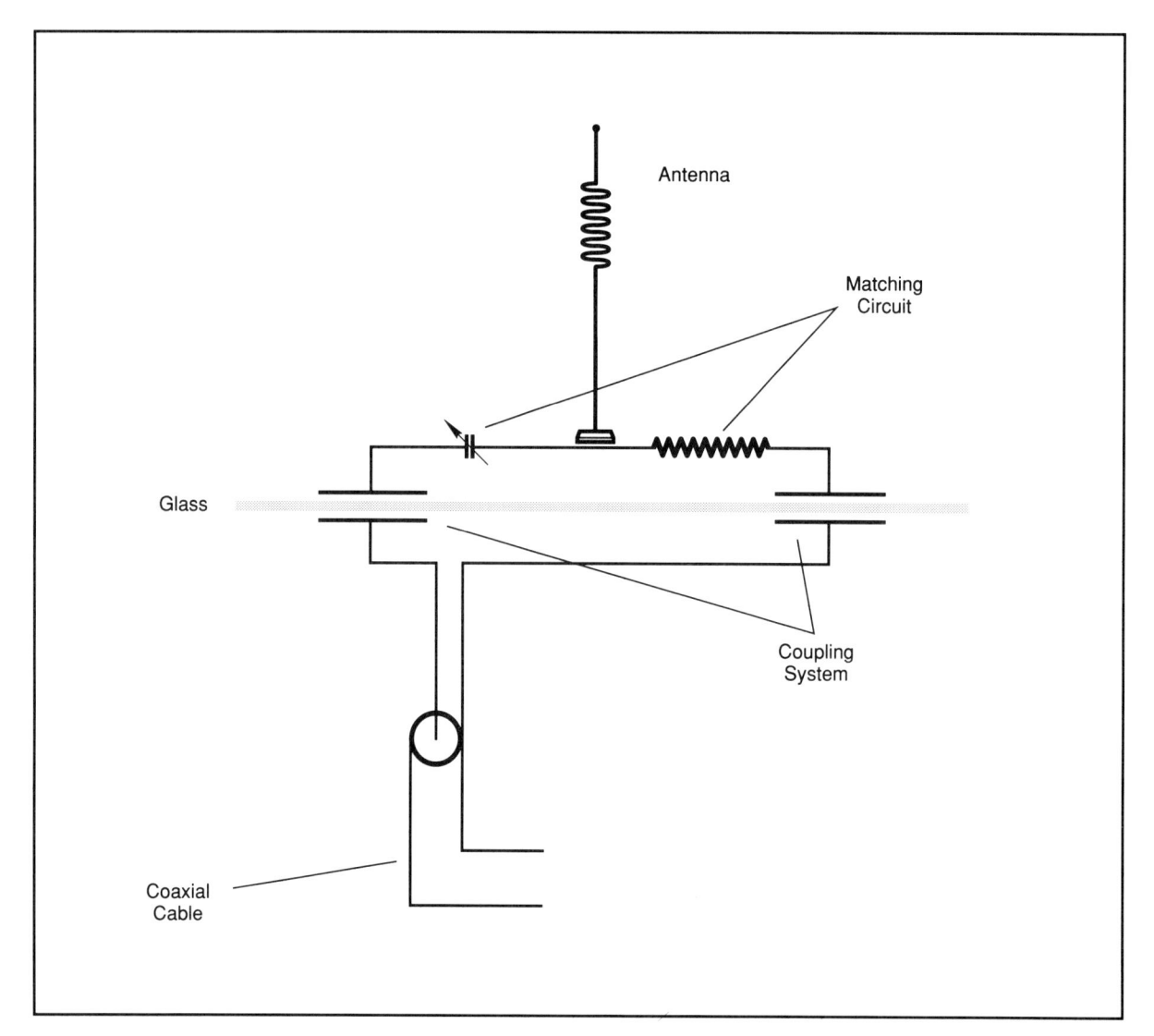

FIGURE 5.11 VARIATION OF THE AVANTE-STYLE GLASS-MOUNT ANTENNA
This is a glass-mount antenna with the matching circuit *outside* the window.

window by using two coupling plates. The low-impedance signal across the outside pair of plates is then transformed up to the antenna feed-point impedance by a matching network mounted outside the vehicle. These manufacturers feel this method offers several benefits to the cellular customer. The first is that the RF energy appearing on the outside plates is at a low impedance and thus reduces the effects of stray coupling to foreign bodies (rain, windshield wipers, defroster wires, etc.). The second is that the electromagnetic fields carried by the transmission line tend to be contained between the plates rather than dispersed in an unconstrained pattern around a single pair of plates. This method of positioning the coupling plates in the transmission line rather than in the antenna avoids antenna loading but results in a longer antenna. While this may not be as cosmetically attractive as the previous design, it does serve to elevate the center point of radiation higher above the roof line and maintain a more omnidirectional pattern of radiation.

Applications. The most obvious application of a glass-mount antenna is for customers who do not want a hole drilled in their car. Although it will not perform as well as a properly installed roof-top or elevated-feed antenna, it does work well when installed properly. Glass-mount antennas are also used by customers who sell their cars every year or where no room exists for any other type of antenna. A very strong supporting argument for glass-mount antennas is their low installation labor requirement. An experienced person can install a glass-mount antenna in less than fifteen minutes with almost no tools.

Elevated-Feed Antennas

The elevated-feed antenna is my favorite choice for a cellular antenna installation (see figure 5.12). It is also the most expensive. A number of manufacturers sell elevated-feed antennas. These products range from cheap and flimsy to high quality and rugged. While I won't recommend any specific one, I suggest that you look at a number of alternatives before selecting a particular manufacturer.

Theory of Operation. Most elevated-feed antennas are five-eighths-wave collinear antennas. Unlike a standard quarter-wave Marconi antenna, they do not require a ground plane. Like a glass-mount antenna, elevated-feed antennas are de-

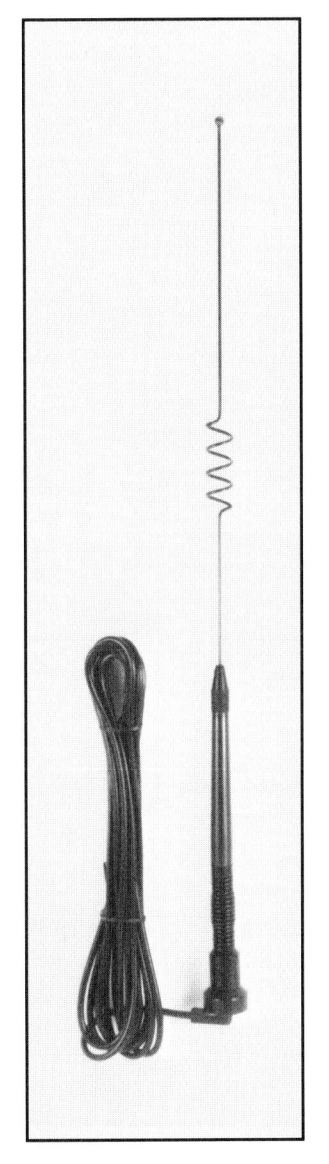

FIGURE 5.12
AN ELEVATED-FEED ANTENNA

signed to provide the user with a certain amount of gain. Most antennas of this type are designed to provide 3 db of gain. A few manufacturers offer 5-db gain models. The installer must take care when selecting antenna gains. In figure 5.13, we can see that the 5-db gain antenna offers more horizontal gain. However, the gain tapers off as the antenna radiates at higher angles.

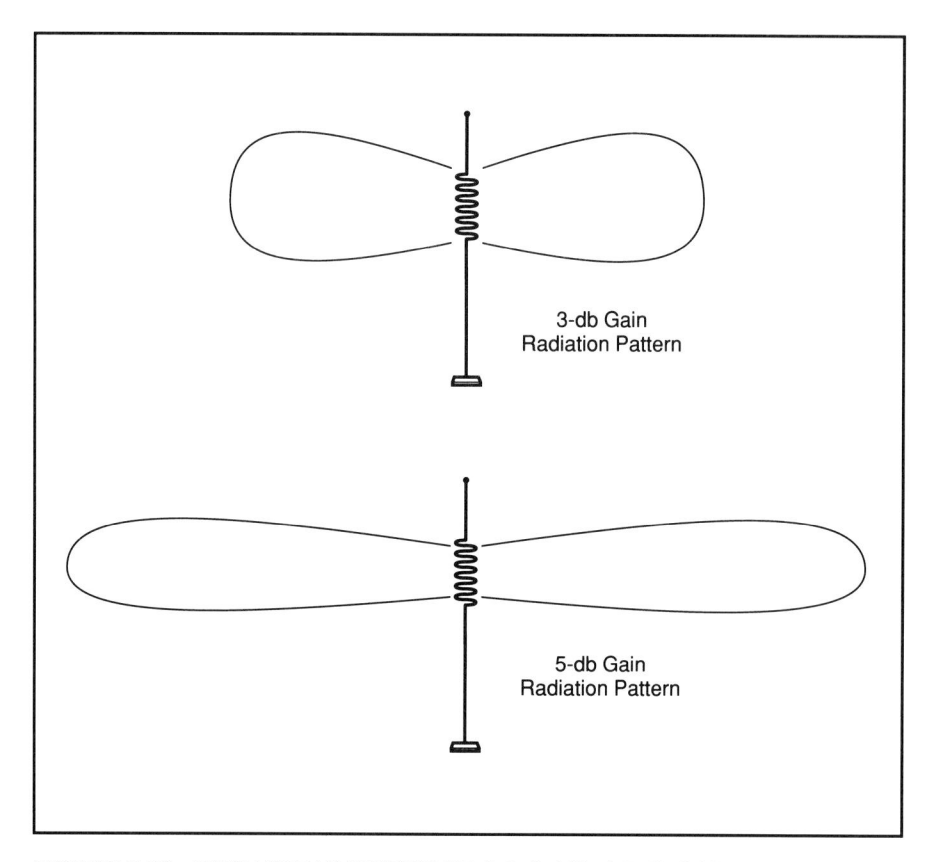

FIGURE 5.13 RADIATION PATTERN OF A 3-db VS. A 5-db GAIN ELEVATED-FEED ANTENNA
Note that the 5-db gain antenna provides more horizontal gain but less vertical gain.

Elevated-feed antennas are rather tall when compared to a standard quarter-wave or five-eighths-wave collinear roof-top antenna. The extra height permits the antenna to be mounted on a trunk lid or fender and still have the radiated signal propagate over the roof of the vehicle.

Applications. The elevated-feed antenna was developed because consumers found a roof-top antenna to be cosmetically unattractive. Because roof-top antennas require a ground plane and are physically small, they cannot be mounted on a vehicle trunk. An installer who attempts this will create a very directional antenna installation. A directional antenna will seriously compromise the cellular telephone's performance. The reason for this is discussed in the last section of this chapter (pp. 96–98).

Elevated-feed antennas are perfect for customers who can be convinced to have a hole drilled in their vehicle but who don't want an antenna on the roof. These antennas also take less time (or at least it seems to me) to install than roof-top antennas.

Roof-Top Antennas

The best performing antenna is also the least popular. Correctly installed, roof-top antennas (see figure 5.14) tend to provide the most omnidirectional patterns of any of the cellular antennas and therefore the best performance.

Theory of Operation. Roof-top antennas are sold in several varieties. The most common are 0-db, 3-db and 5-db gain antennas (see figure 5.15). Most are a variation of the Marconi-type antenna that was described earlier in this chapter. Roof-top antennas require a ground plane in order to achieve maximum gain. The ground plane acts as a mirror image to the quarter-wave antenna mounted on the roof. Because the ground plane affects the antenna gain, it is important to have sufficient area to reflect the signal. About 3 inches on all sides is usually sufficient at 800 MHz. Lower frequencies require more ground plane.

There is considerable debate over how many decibels of gain one should have in a roof-top antenna. More gain is not necessarily better. As the gain of an antenna is increased, its ability to transmit and receive signals decreases as the source becomes closer to vertical. In figure 5.16, the 3-db gain antenna can actually communicate with the cell site on top of the mountain better than the 5-db gain antenna. This is because the 5-db gain antenna gives up its ability to transmit and receive signals the closer it is to vertical (with respect to the antenna).

There are a few good rules that can be followed when selecting a roof-top antenna.

1. Generally, 0-db gain antennas are not recommended. They are to be used only in situations where a short antenna is an absolute must, such as for a truck that is close to the

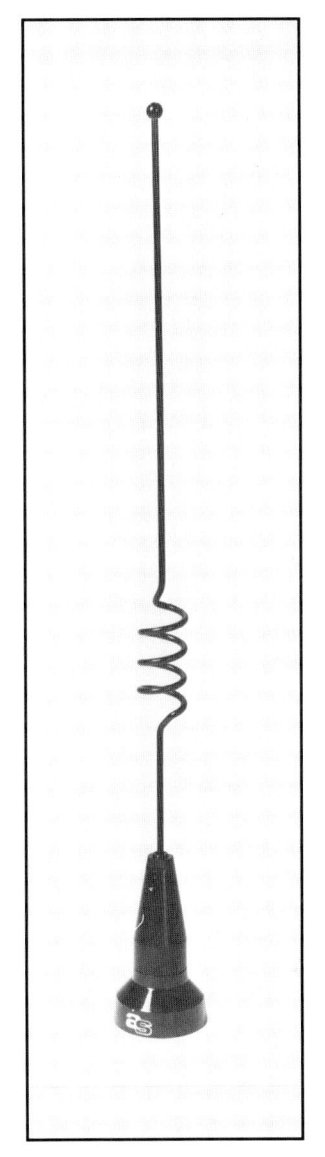

FIGURE 5.14
A ROOF-TOP ANTENNA

94 ANTENNA THEORY AND SELECTION

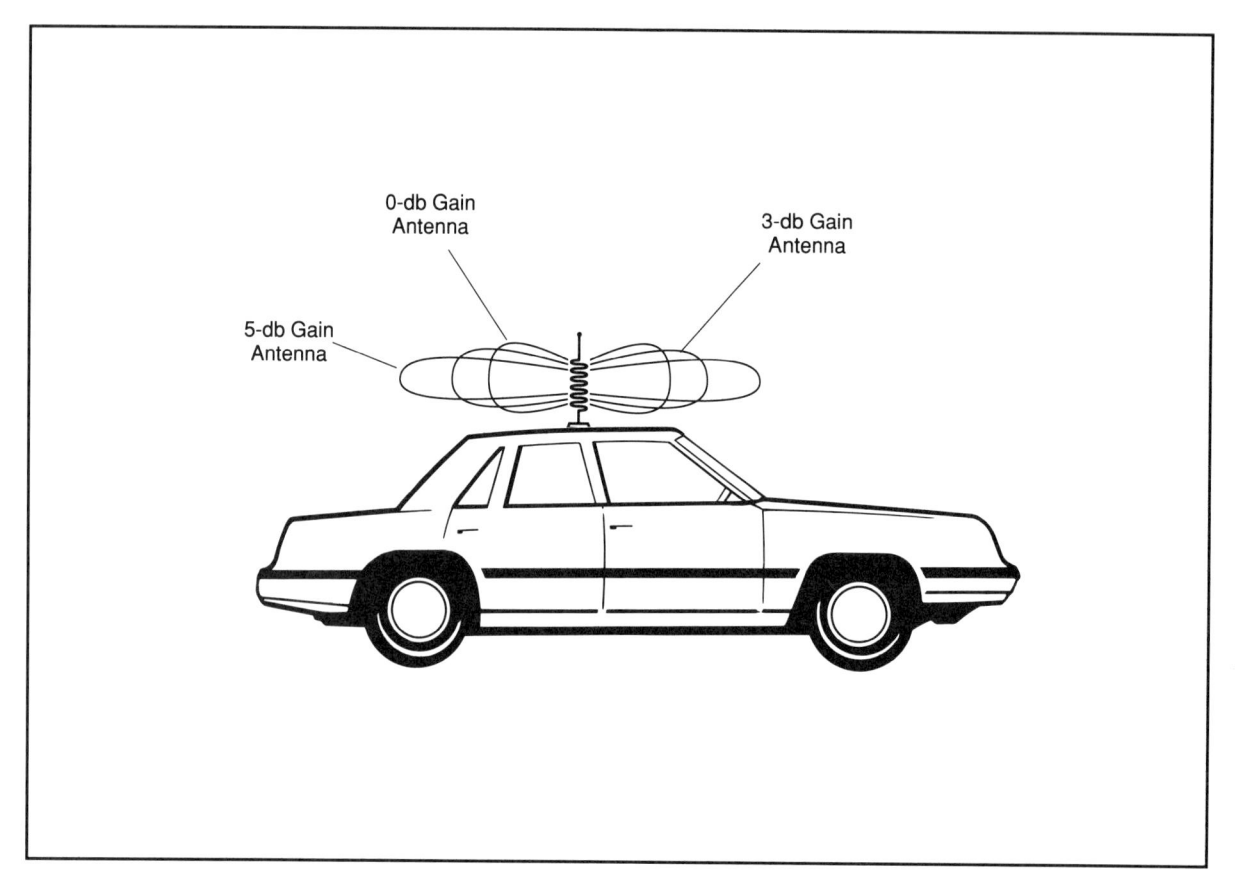

FIGURE 5.15 ROOF-TOP ANTENNAS WITH VARIOUS GAIN
The greater the horizontal gain of an antenna, the lower the vertical gain.

maximum height for a garage or tunnel. In these situations, it is okay to use a 0-db gain antenna because the antenna is physically quite high; 0 db of gain is always better than a broken antenna. Another application is for the customer who wants a small, unobtrusive antenna.

2. The most common roof-top antennas are 3-db gain antennas. They are used in situations where the customer wants the best possible performance and is not too concerned about drilling a hole in the roof. They provide a bit of extra gain but don't have radiation patterns that are too flat. They will work in both urban and rural areas where the terrain is either hilly or flat. In the "concrete canyons" of New York City, a properly installed 3-db gain antenna is the antenna of choice.

ANTENNA THEORY AND SELECTION 95

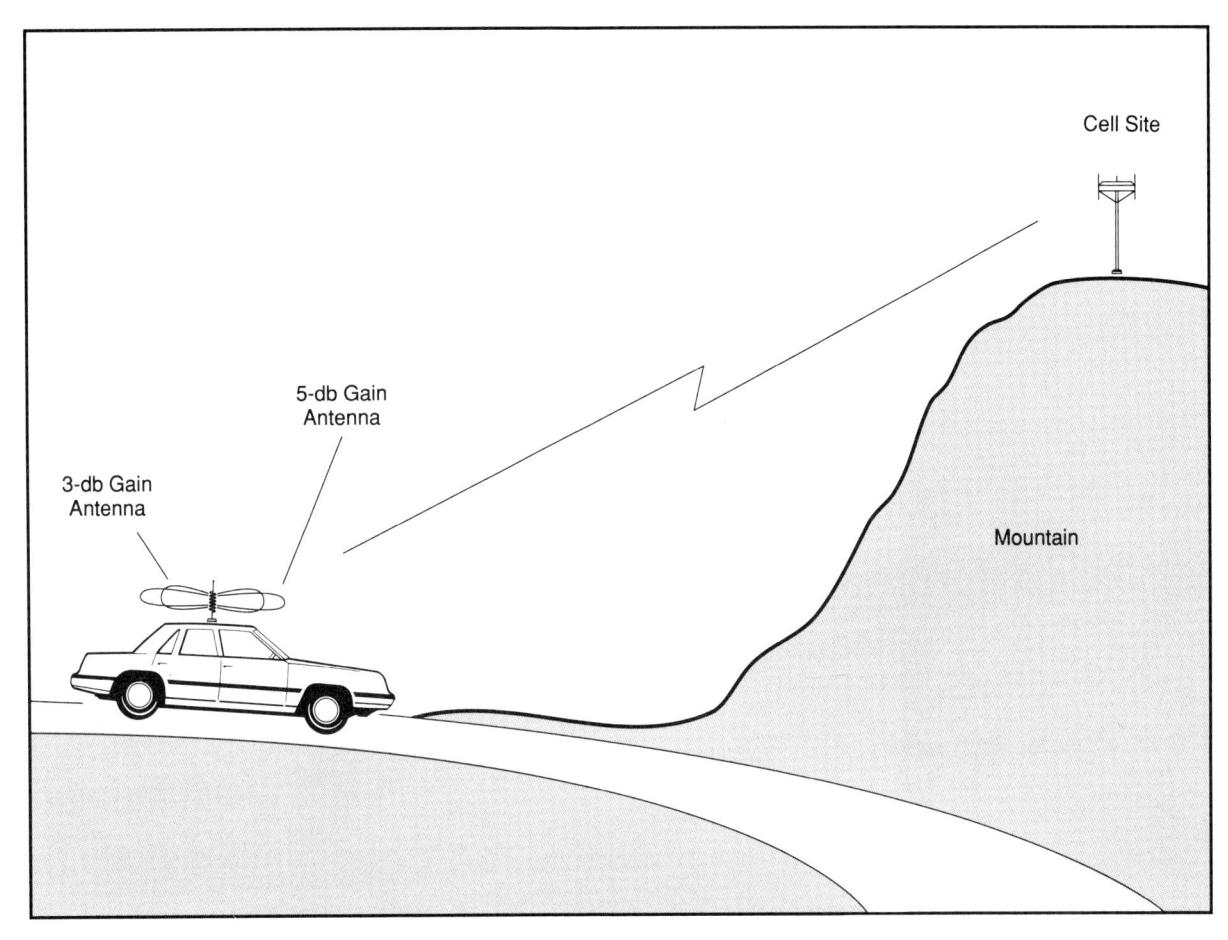

FIGURE 5.16 RADIATION PATTERN OF A 3-db VS. A 5-db GAIN ROOF-TOP ANTENNA
A 5-db gain antenna will not work well when communicating with a cell site at a higher elevation because it lacks sufficient vertical gain.

3. The 5-db gain antennas are a bit tricky. While I am sure that some of the manufacturers of these antennas would disagree with me, I don't think they are suitable for many urban applications. In order to get some extra gain on the horizontal plane, these antennas are deficient in gain as one approaches vertical. Since urban areas are plagued with multipath interference and other RF headaches, these antennas only serve to complicate matters. In areas with such problems, they can contribute to additional handoffs and dropped calls. However, in areas with flat topography, they can considerably improve reception. I would recommend them in rural areas where the coverage is sparse.

Applications. Sedans and trucks are the best choices for roof-top antennas. People who are proponents of function over style are candidates for these antennas. When I worked for an installation facility, we installed most of our roof-top antennas on company cars and American sedans. These cars were driven by people who wanted to get the maximum performance out of their phones. Large trucks are also good candidates. The drivers are not concerned about a hole in the roof of their cab—they simply want the best performance. The 5-db gain antennas are useful when installed in long-haul trucks. Since these trucks are driven in a lot of marginal coverage areas, the 5-db gain antenna can sometimes make the difference.

THE LOSEE-SHOSTECK EFFECT

A few years ago, while working with a Washington, D.C.-based consultant named Herschel Shosteck, we discovered a relationship between improper installations and dropped calls. Someone in the trade press called it the Losee-Shosteck effect. This phenomenon is quite simple to understand once a few facts are known.

1. Improperly installed antennas are directional.
2. The highest risk of dropping a call occurs during the handoff process.
3. Directional antennas in a mobile environment hand off more than nondirectional antennas.

This last item requires a bit of explanation. In figure 5.17, a cellular phone has been improperly installed: a roof-top antenna was mounted on the fender of the vehicle. The antenna becomes highly directional because the vehicle cabin blocks almost all forward signal propagation and the uneven ground plane further distorts the pattern (see figure 5.18). Because the antenna transmits and receives more strongly in one direction, it communicates with cell site 2 instead of cell site 1, which it is actually closer to. As the vehicle moves along the road, the call is handed off to cell site 2, then eventually to cell site 1. Had the antenna been nondirectional, it would have been operating on cell site 1 the entire time. The directional antenna caused two unnecessary handoffs.

This theory was backed up by observations made in New York City. It was found that directional antennas in urban

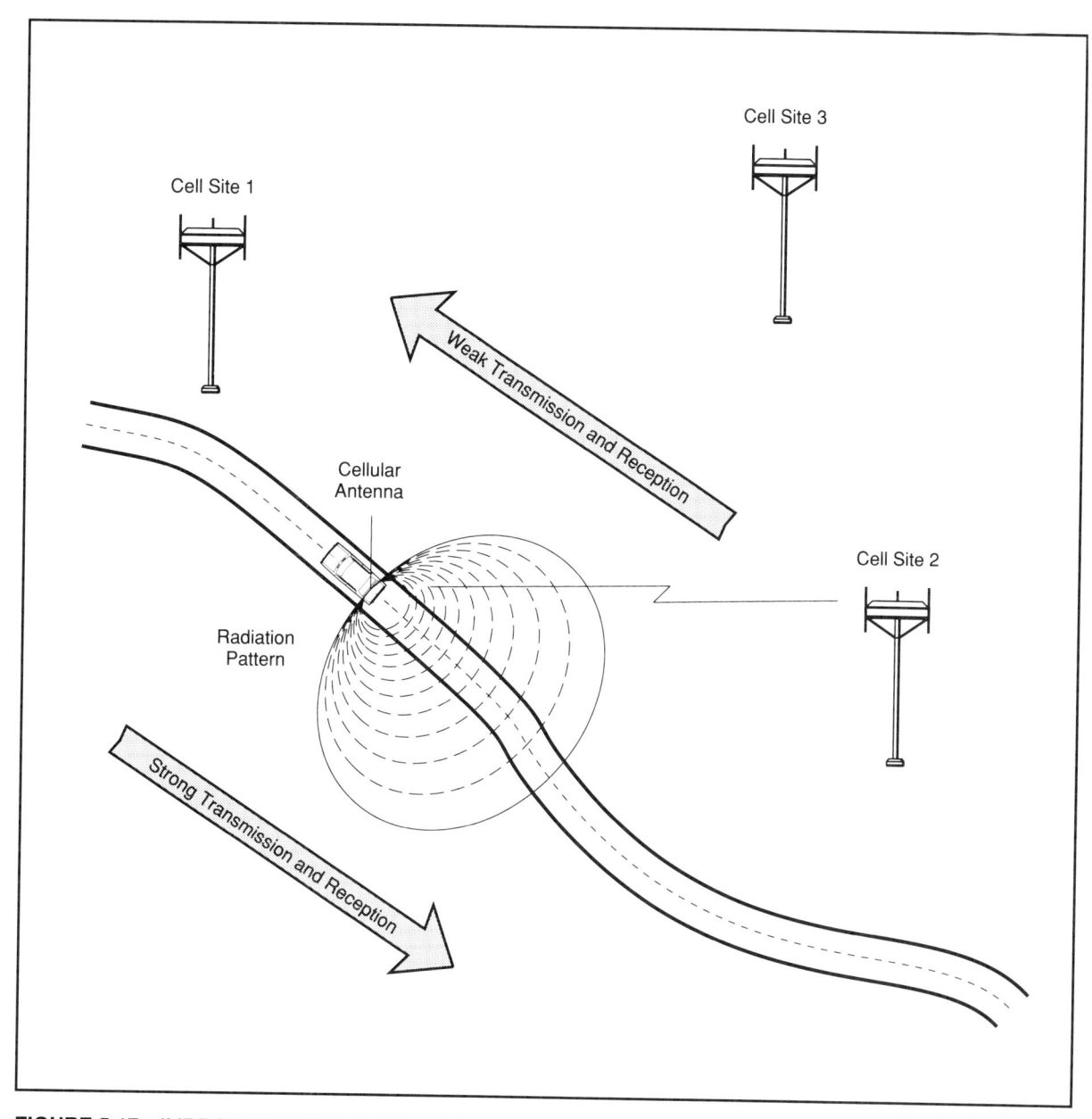

FIGURE 5.17 IMPROPERLY INSTALLED ROOF-TOP ANTENNA
The antenna communicates with the "wrong" cell site because it is directional.

environments such as New York City increased handoffs. As a vehicle moves, its physical relationship to a cell site changes constantly. If the gain of the antenna is not the same in every direction, the cell site can lose SAT, receive a poor signal, or make a data error during the handoff process. The result can be a

98 ANTENNA THEORY AND SELECTION

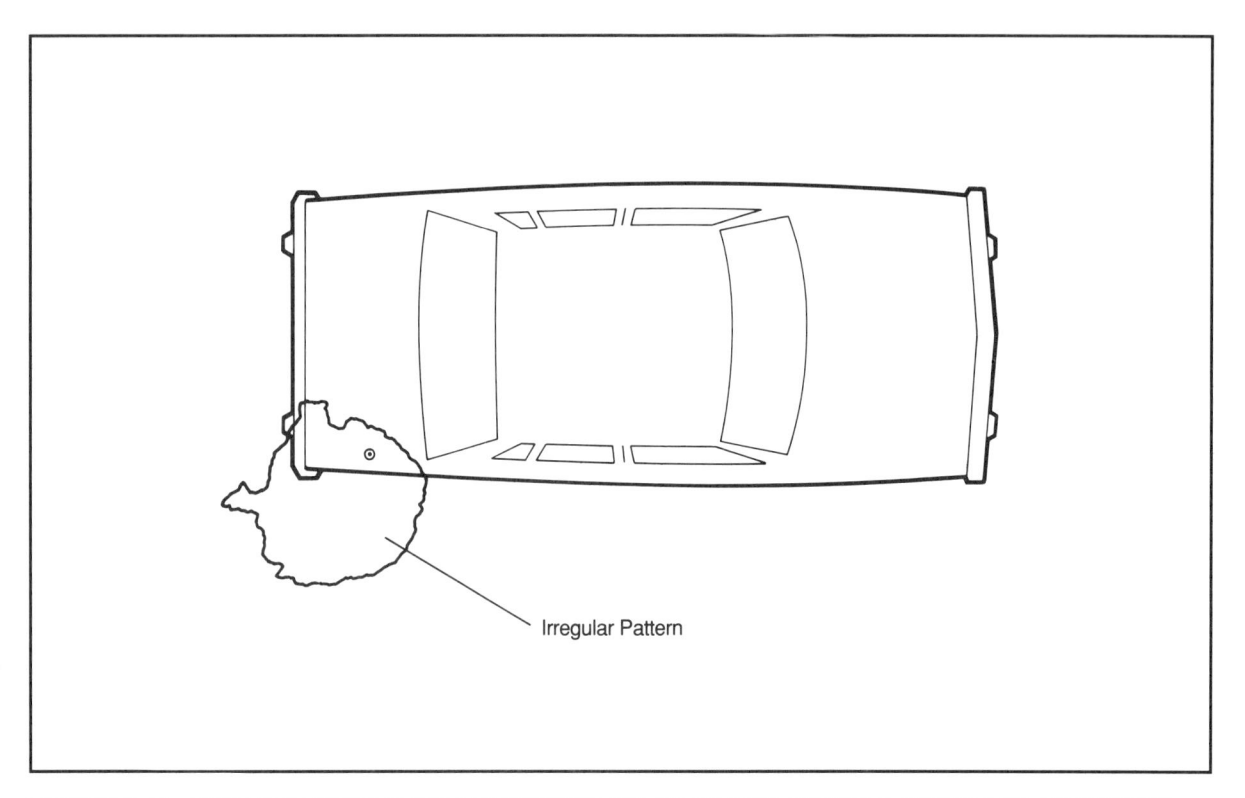

FIGURE 5.18 TOP VIEW OF RADIATION PATTERN FOR ROOF-TOP ANTENNA INSTALLED ON FENDER

dropped call or a handoff. As stated earlier, every time a handoff occurs, there is a chance that the call will be lost due to data errors or be handed off to a cell with no available channels. If the phone is handed off enough times, the call will be dropped.

Increased handoffs also put an additional load on the cellular system. With some of the larger systems already on the verge of overload, the last thing a system operator wants is to process additional handoffs. Most carriers are now monitoring handoffs and generating lists of customers who drop too many calls. The carriers then contact these customers and suggest that they have their cellular unit checked by the people who installed it.

The moral of this story is to use the proper antenna in each installation. The wrong antenna will aggravate the customer, the carrier, and you.

CHAPTER 6

Vehicle Installations

After spending the time to learn some of the theories and concepts that make cellular work, you are now ready to learn how to install a cellular telephone in a vehicle. This is the fun part of the business—and the part where the installer and facility owner make some money.

PRECHECKING THE VEHICLE

While the customer is taking care of the financial part of the transaction or is discussing the particular type of phone that he/she wants, the service manager should check out the customer's car. The purpose of checking out the car is to gather information that will be useful during and after the installation. As explained in Chapter 3 (pp. 51–52), it is a good idea to use a standard form to collect this information. The form should be the same one that is used to enter the information into the computer. By doing this, clerks will have one less piece of paperwork to file (or lose). A good starting point for an installation work record is shown on the next page.

After this information has been collected, the service manager should make a complete mechanical check of the vehicle. Sometimes customers are not aware of existing problems with their vehicles. On the other hand, there are always people who are looking for something for nothing or who just aren't very honest. People like this are the exception to the rule, but it only takes a few of them to create problems for the service manager and

100 VEHICLE INSTALLATIONS

```
                        INSTALLATION WORK RECORD

COMPANY NAME: _____

CONTACT:   LAST NAME: _____  FIRST NAME: _____

ADDRESS: _____  OFFICE PHONE NO.: _____

CITY: _____  STATE: _____  ZIP: _____

CELLULAR PHONE NO.: (    ) _____  ESN: _____

PHONE MANUFACTURER: _____  MODEL: _____

TRANSCEIVER SERIAL NO.: _____  CONTROL HEAD SERIAL NO.: _____

INSTALLER'S INITIALS: _____  IN-TIME: _____  OUT-TIME: _____

VEHICLE MAKE: _____  MODEL: _____  YEAR: _____

LICENSE PLATE: _____  STATE: _____

SALESPERSON: _____  INSTALLED PRICE: _____

CONTROL HEAD LOCATION: _____

TRANSCEIVER LOCATION: _____
```

installer. An example of such a situation is a customer with a defective car radio. It would be very easy to blame the problem on the cellular installation facility if the car had not been checked out before the installation. Take the time to do a complete electrical and mechanical check of the vehicle. With a good checklist, the process takes only a few minutes. It can save you from being blamed for an error that you didn't make. On the following page is a list of items that can be quickly checked BEFORE beginning a cellular phone installation.

After completing this checklist, be sure the customer reads and signs the list. Customers should understand that by signing the list they are releasing the installer from responsibility for any

```
                    PREINSTALLATION CHECKLIST

CAR STARTS: _____

HEADLIGHTS: _____

BRAKELIGHTS: _____

TAILLIGHTS: _____

DIRECTIONAL SIGNALS: _____

DRIVING LIGHTS: _____

DOME LIGHT: _____

DASH LIGHTS: _____

STEREO: _____

GAUGES (GAS, OIL, ETC.): _____

ELECTRICAL ACCESSORIES (SEAT, WINDOWS, LOCKS): _____

ALARM SYSTEM: _____

ELECTRIC SUNROOF OPERATION: _____

RADAR DETECTOR: _____

DENTS AND SCRATCHES: _____
```

existing mechanical or electrical problems. This should eliminate any chance of a customer blaming these problems on the installer.

If there is an alarm system in the car, you will need to get some additional information from the customer. It can be rather embarrassing to trip the alarm (which will happen if it is not disabled) during the installation. These alarms make a great deal of noise and can be rather difficult to disable if you don't know how to do it. There are a number of vehicle alarms available, ranging from simple to complex and some costing over $1,000. They are very sophisticated and difficult to disarm if you don't know how to. Have the customer give you a complete description of the alarm and its disarm code. After awhile, you will become familiar with the different types of alarms and will be able to understand them quickly. Learning how to disarm them will save everyone a great deal of trouble.

Before planning the installation, suggest that the customer remove any valuables from the car (such as money, radar detectors, Walkman, briefcase, etc.). Once again, this eliminates a potential problem. It should also make the customer feel confident that he/she is doing business with a company concerned about integrity.

PLANNING THE INSTALLATION

Once the customer has approved the checklist, it is time to discuss the installation. The installation should never be planned by a salesperson—only by a service manager. I am not questioning the integrity of salespeople, but rather stating that it is much easier for someone to promise something to a customer when somebody else is responsible for fulfilling that promise.

Selecting the Control Head Location

The first item that the service manager should discuss with the customer is placement of the control head. This is one of the most important aspects of the installation because it is the part of the phone with which the customer will have the most contact.

Figure 6.1 shows a common location for the control head. It is easily mounted on the console between the front seats. The user can quickly reach the phone, and almost no cable will be visible after the installation. This location is advantageous to both the installer and the customer: it is the easiest mounting location, and it doesn't require any holes to be drilled in the car.

As shown in figure 6.2, another good location to mount the control head is on the dashboard. Some customers favor this approach because it makes the control head readily accessible. This mounting is easier to do in some cars than in others. For example, American cars usually have flat surfaces in the dashboard that allow the mounting hardware to be attached easily.

VEHICLE INSTALLATIONS 103

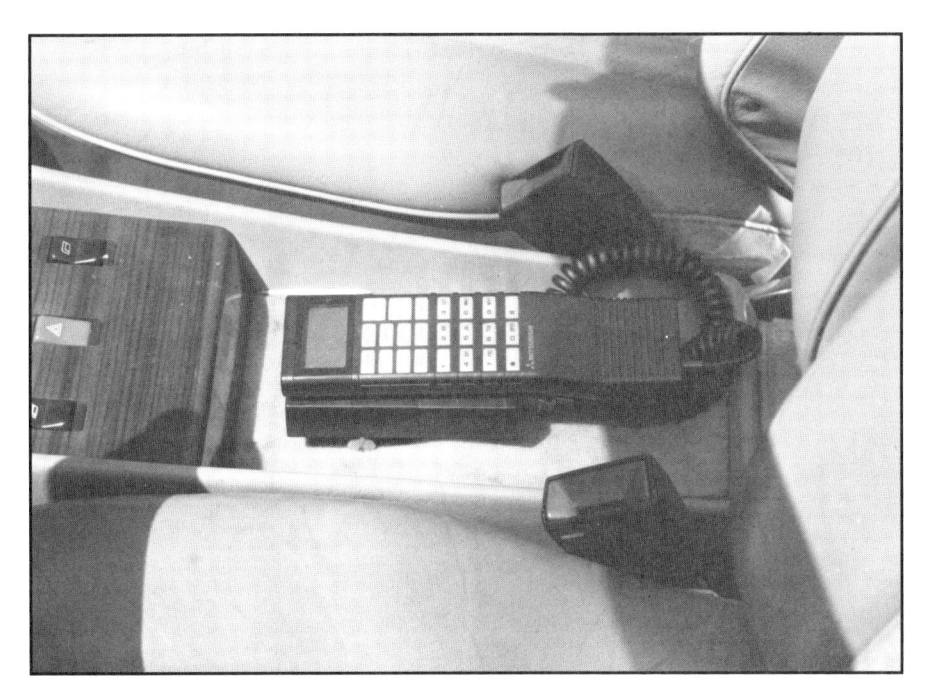

FIGURE 6.1 CONTROL HEAD MOUNTED ON CENTER CONSOLE

The dash mounts, while providing easy access to the control head, require that the installer drill holes in the dash, and many customers find this objectionable. Dash mounts can also present routing problems for the handset cable.

The most popular mounting location in sedan-type vehicles is on the floor, as shown in figure 6.3. The control head is mounted on a pedestal mount on the center console. This once again provides easy access to the phone. It also offers the advantage of being easy to cover with a coat when the car is parked. This location is easy to install, and the holes in the floor are not noticeable (because the carpet conceals them) if the phone is removed at a later date.

Selecting the Transceiver Location

Once the location of the control head is decided, the transceiver location must be discussed. The location depends upon a number of different factors. The type of vehicle and cellular phone both play an important role in this decision. There are four basic types of transceiver mounting situations:

FIGURE 6.2 CONTROL HEAD MOUNTED ON DASHBOARD

FIGURE 6.3 CONTROL HEAD MOUNTED ON FLOOR

1. Sedan (vehicle with a trunk) and mobile-only phone
2. Sedan and transportable phone
3. Hatchback and mobile-only phone
4. Hatchback and transportable phone

The sedan with a mobile-only phone offers the installer and the customer the most flexibility. The service manager should ask the customer if he/she plans to transfer the telephone from car to car. Many people want to have this flexibility if they have more than one car. This can be accomplished by putting an installation kit in each car. The control head and transceiver are then removed by the customer and transferred from car to car. If the customer wishes to have this capability, the transceiver must be mounted in a location where the customer can get to it easily. Figure 6.4 shows a mobile-only transceiver mounted in a convenient location in the trunk of a sedan. Note that the transceiver has been mounted so that the power and antenna cables can be accessed easily by the customer when he/she wants to transfer the phone from car to car. The service manager should suggest a similar location for a transportable cellular phone.

FIGURE 6.4 MOBILE-ONLY TRANSCEIVER MOUNTED IN TRUNK OF SEDAN (REMOVABLE)

If the customer plans to operate his/her cellular telephone in only one vehicle, the location possibilities are much more varied. As cellular transceivers become smaller, the installer has more choices when selecting transceiver locations.

Trunk Location. My favorite location for the transceiver is in the trunk. If the customer doesn't need to remove the phone to put it in another car, the transceiver can be mounted out of the way, such as in one of the rear corners of the trunk (see figure 6.5). In this photo, you will notice that the cables are tucked out of the way so that they are not conspicuous and also will not be damaged by luggage or cargo in the trunk.

Mounting the transceiver in the trunk offers a few advantages. The first is that it keeps the antenna cable short. Elevated-feed antennas are mounted on the rear fender of the vehicle, so if the transceiver is mounted on the same side as the antenna, the cable will be only a couple of feet long. This keeps the signal loss to a minimum. A trunk-mounted transceiver is also easy to get to in the event it needs service at a later date.

FIGURE 6.5 MOBILE-ONLY TRANSCEIVER MOUNTED IN CORNER OF TRUNK

Driver's Seat Location. Another ideal location for the transceiver is under the driver's seat (see figure 6.6). This isn't possible in some vehicles; most vehicles with electric seats don't have enough room. The vast majority of cars, however, do have enough space for the installer to mount the smaller transceivers that are being manufactured today.

Special care should be taken when using this approach in colder climates. During the winter, salt is frequently used by snow removal crews to melt ice on the road. The salt can get on the shoes of the driver and then work its way into the vehicle carpet. If the salt gets on the transceiver or on the contacts, it will corrode the metal and damage the transceiver. Be sure there is no way the transceiver will be damaged by salt from the driver's shoes. If there is any doubt, mount the transceiver in the trunk. *Remember:* Warranties don't cover salt damage.

There are a few significant advantages to mounting the transceiver under the driver's seat. The first is ease of installation. The power wires are kept short, and the transceiver can be easily mounted with velcro. There are no holes drilled, and the customer doesn't have to worry about damaging the transceiver with luggage in the trunk. In trucks, this is about the only location

FIGURE 6.6 TRANSCEIVER MOUNTED UNDER DRIVER'S SEAT

that makes any sense. Under-seat mounting is not a good idea for a customer who purchases a transportable telephone. There is usually not enough room to remove and install the transceiver easily.

Hatchback Location. Hatchbacks present a different set of problems. With these vehicles, I usually suggest mounting the transceiver under the seat whenever possible. However, when the seat configuration or customer preference will not permit it, the back deck is the only answer. Transportable units must also be mounted on the rear deck. There is not one particular place that is a universal best choice. Figure 6.7 shows a transceiver mounted on the back deck of a hatchback. But the installer must decide with the customer the best location for the transceiver in that particular car.

Selecting the Antenna Location

Glass-Mount Antennas. When installing a glass-mount antenna, location is of primary importance. My first choice is always on the rear window of the vehicle (see figure 6.8). Mounting on any of the side windows breaks up the lines of the car, and

FIGURE 6.7 TRANSCEIVER MOUNTED ON BACK DECK OF HATCHBACK

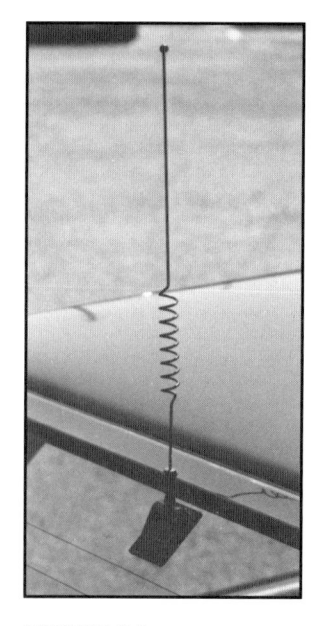

FIGURE 6.8
GLASS-MOUNT ANTENNA MOUNTED ON REAR WINDOW

it is difficult to hide the cable. If the vehicle is a convertible, the only location for a glass-mount antenna is on the front windshield, usually in the center or in one of the corners. In convertible installations, I suggest an elevated-feed antenna. It will usually outperform and look better than a glass-mount antenna.

Mounting on the back window is usually possible and is the best choice. Select a location as close to the top of the window as possible. By locating the antenna at the highest point, the rear-window defoggers are avoided, and the signal radiated from the antenna is able to clear the top of the vehicle (see figure 6.9). Rear-window defoggers is an area of some controversy with regard to glass-mount antennas. It is best to avoid mounting the coupling box near or on top of the defogger wires. If the box is mounted in close proximity to these wires, some power from the transceiver is coupled into the wires. Most of the power coupled into the defogger wires is then radiated as if the wires were an antenna. This tends to make the installation somewhat directional, and in mature urban markets, the result can be an increased rate of dropped calls. This was discussed earlier in Chapter 5, pp. 96–98.

The worst installation choice is to mount the coupling box directly over one defogger wire (see figure 6.10). In some vehicles, the installer will have no choice but to mount the coupling box over the defogger wires. A good compromise is to mount the

FIGURE 6.9 GLASS-MOUNT ANTENNA LOCATED NEAR TOP OF WINDOW
Proper positioning of the glass-mount antenna allows the signal to clear the top of the vehicle, ensuring an omnidirectional radiation pattern.

coupling box so it "straddles" two defogger wires (see figure 6.11). The absorbed power from the transceiver tends to be cancelled by the two defogger wires. The result is almost no lost power and very little pattern distortion.

If a vehicle appears to have no defogger wires, a closer look will most likely reveal a fine wire mesh that heats the entire rear window. While this makes a very nice defogger, the glass-mount antenna will not work. Mercedes-Benz uses this approach in some of their vehicles. In this situation, the elevated-feed antenna is the next best choice.

Another important item to consider when choosing a mounting site is proximity to a windshield wiper. Some glass-mount antennas (high-impedance coupler design) can be affected by the metal in the wiper blades. A good rule is not to mount the coupling box where it is closer than 1/2 inch to the wiper blade path.

Aftermarket window tinting can also create problems for glass-mount antennas. Most tinting materials contain some metal. This will seriously detune the antenna and significantly degrade the performance of the cellular telephone. Most aftermarket

110 VEHICLE INSTALLATIONS

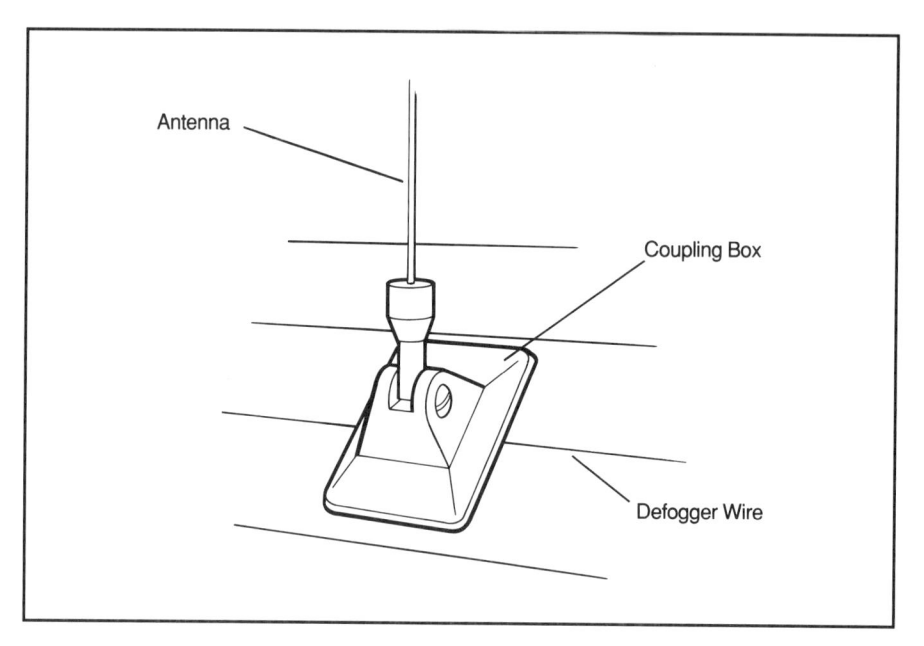

FIGURE 6.10 IMPROPER COUPLING-BOX INSTALLATION
Mounting the coupling box over one defogger wire causes the signal to be coupled into the defogger wire instead of the antenna.

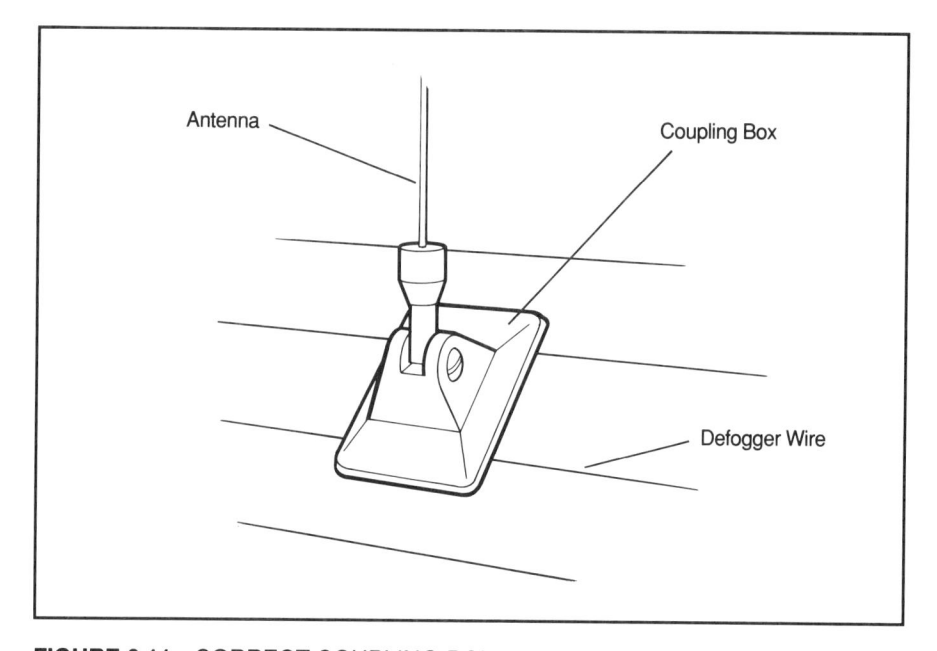

FIGURE 6.11 CORRECT COUPLING-BOX INSTALLATION
Mounting the coupling box so that it "straddles" two defogger wires ensures that maximum signal is delivered to the antenna.

window tinting can be removed with a sharp knife. If the customer does not object (be sure to check first, since it is a lot easier to remove the material than to put it back), trace the coupling box pattern on the window tinting material with a very sharp razor knife. It is then quite easy to peel the tint off the window with a fingernail. Be sure the window is cleaned thoroughly before attempting to mount the coupling box.

Some Rolls-Royces can present problems for installers wishing to use a glass-mount antenna. Some models contain lead in their windows. Lead is used to make high-quality glass more transparent. Unfortunately, it is not possible to use a glass-mount antenna with this type of glass because the lead absorbs a great deal of the power from the transmitter. An installer who thinks that a particular Rolls-Royce has this type of glass should consult with the manufacturer. In this situation, the best choice is to convince the customer to use an elevated-feed antenna. Tell the customer that real men (and women) are not afraid to drill holes in their cars.

Elevated-Feed Antennas. Location selection is easy for elevated-feed antennas. As shown in figure 6.12, mount the antenna on either rear fender, usually across from the vehicle AM/FM antenna (for cosmetic reasons). The antenna can be mounted in the center of the trunk, but the installation is a bit more difficult and does not look as nice as when the antenna is mounted on the fender. The only serious caution that applies is to look around the site where you plan to drill the hole. Be sure that there is no way the gas tank can be punctured or that a wiring harness can be damaged. Late model cars, particularly expensive foreign models like BMW, Mercedes-Benz, Porsche, or exotics, have gas tanks and wiring harnesses in strange places. It makes sense to be sure nothing will be damaged any time a hole is to be drilled.

When selecting the antenna mounting location, be sure that at least 2 inches of mounting surface is available. The surface must be flat with no ridges or other surface features that will prevent the antenna from making a good mechanical contact. If the antenna is not securely mounted to the trunk or fender surface, water can leak into the trunk through any space in the seal. The antenna should also be mounted as close to vertical as possible. Remember, elevated-feed antennas are horizontally polarized, so any significant deviation from vertical (more than a

FIGURE 6.12
ELEVATED-FEED ANTENNA MOUNTED ON REAR FENDER

112 VEHICLE INSTALLATIONS

few degrees) will cause the antenna to become directional. And directional antennas are the cause of a myriad of cellular terminal performance problems.

Roof-Top Antennas. Rule number 1 for rooftop antennas: INSTALL THEM ON THE ROOF AND ONLY ON THE ROOF OF THE VEHICLE (see figure 6.13).
Rule number 2: SEE RULE NUMBER 1!!!

The easiest way to "ruin" a cellular telephone installation is to mount a roof-top antenna on the vehicle trunk or fender. When this is done, the antenna becomes very directional and will drop two to three times as many calls as a properly installed antenna. Figure 6.14 shows the radiation pattern of a roof-top antenna that has been mounted on a rear fender. It becomes very directional because the vehicle cabin blocks almost all forward signal propagation and the uneven ground plane further distorts the pattern. The reasons for this were explained in the previous chapter (see pp. 96–98). *Remember: Roof-top antennas go on the roof.*

FIGURE 6.13 ROOF-TOP ANTENNA MOUNTED ON ROOF

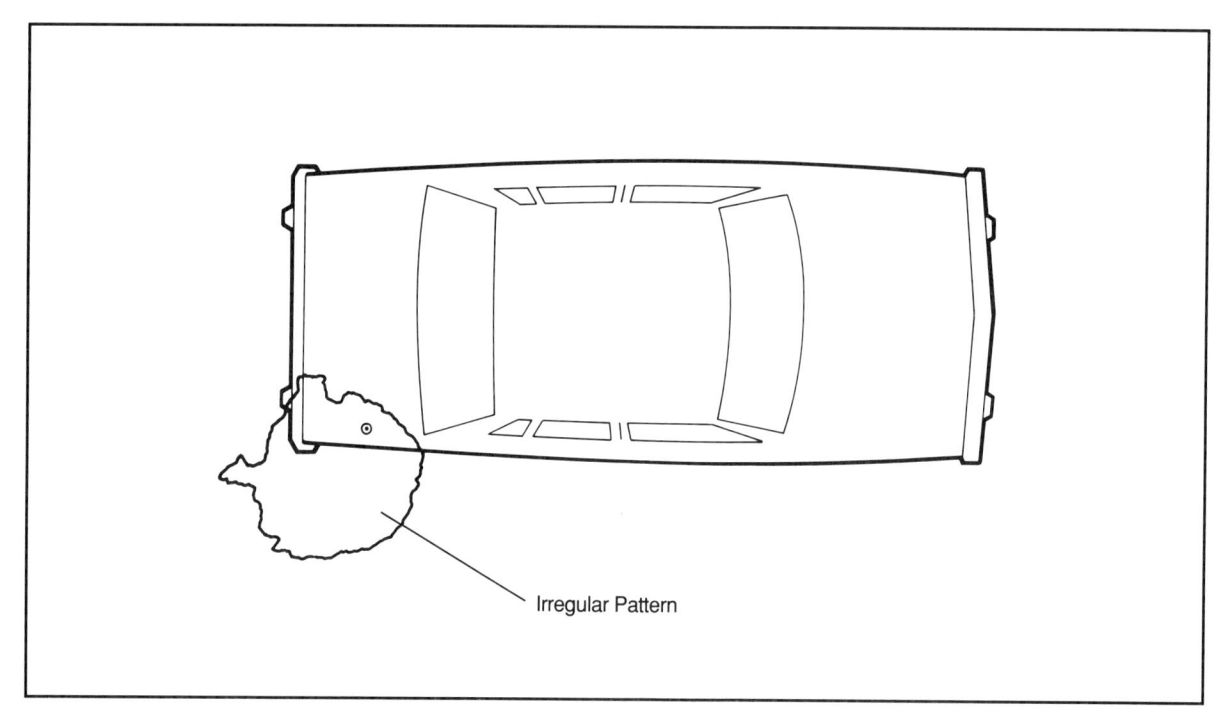

FIGURE 6.14 TOP VIEW OF RADIATION PATTERN FOR ROOF-TOP ANTENNA MOUNTED ON REAR FENDER

The best place to mount the antenna is in the center of the roof. By locating the antenna there, the radiation pattern will be very close to omnidirectional. However, this location is not always possible—automobile manufacturers have a habit of selling cars with sunroofs, ski racks, and decorative trim in the exact place where the antenna would ideally be located. Here are a few suggestions for installation locations:

1. OVER THE DOME LIGHT. Most vehicles have a dome light on the interior of the roof. Directly above this light is an ideal location for the antenna. The light is usually in the middle of the roof and far enough back from the front of the car so that there will be almost no directionality (from an uneven ground plane). The dome light is normally easy to remove. Once the light fixture has been removed, the hole is very easy to drill.

2. REAR CENTER. If the vehicle is equipped with a ski rack or sunroof, the rear center of the roof is a good location for the antenna. Be sure the antenna is at least 3 inches from the rear edge of the roof. If the vehicle is equipped with a retracting sunroof (one that slides into the rear portion of the roof), a roof-top antenna is not suggested. In this case, the rear portion of the roof is usually filled with motors, gears, and cables to retract the sunroof. Drilling a hole in a sunroof motor is a great way to lose the profit from the next ten installations.

3. CONVERTIBLES. This is the only situation where it is permissible to mount a roof-top antenna on the trunk. Because there is no metal cabin roof to block the signal, this is an acceptable location as long as the antenna is mounted in the center of the trunk.

4. CORVETTES. Installations in these cars can be difficult. The owners can be even more difficult. They are usually very leery about having holes drilled in their "babies." The best place to mount an antenna (particularly in convertibles) is on the gas cap. It is the only piece of metal in the body of the car. Metal is a lot easier to drill than fiberglass, and it also provides a ground plane. The gas tank underneath also helps. The installation looks sharp, and a gas cap is inexpensive to replace if the owner gets upset about the hole.

BEGINNING THE INSTALLATION

We are at the point where we need to make our first decision concerning the installation—to disconnect or not to disconnect the car battery.

Security Radios

Many of today's high-end automobiles are equipped with "security radios." These security, or antitheft, radios are equipped with microprocessors that monitor the vehicle voltage. When a thief steals one of these radios, the power is interrupted when the cables are cut or unplugged. The radio's internal microprocessor recognizes this and disables the radio. Even after the power is restored, the unit will not work. Unless the proper "unlock" code is entered into the radio, it becomes an expensive paperweight. To further complicate the problem, many customers don't even know that their car radios are equipped with such a feature. It helps to ask but, *when in doubt, don't disconnect the vehicle's battery*.

Below is a list of vehicles that may be equipped with antitheft radios. It is important to remember that not all of the vehicles on the list are equipped with these radios, and I'm sure there are a few that I have left out.

1. Mercedes-Benz
2. Volvo
3. Saab
4. Jaguar
5. BMW
6. Other high-end European vehicles

If you are unfortunate enough to "lock up" a radio, there are several courses of action you can take.

1. First, ask the customer if he/she knows how to unlock the radio. If they have ever had to "jump" start the car or if the battery has gone dead, they will probably know how to reactivate the radio.
2. If the customer doesn't have this information, check the glove compartment and look for the vehicle operator's manual. The radio manufacturer will often put the unlock code in this manual. Be sure to ask customers if they object to you looking through their glove compartment.

3. If that doesn't work, call the dealer where the customer purchased the car. Explain to them what has happened. They may be unwilling to provide the information over the phone, but if you bring the locked radio back to their facility, they will almost always help. The customer may even be willing to stop by the car dealer on the way home from your facility. It is a good idea to develop a working relationship with the local car dealers, since it is likely that you will "lock up" more than one car radio during your installation career.

On-Board Computers

One more precaution about disconnecting the battery. Many of the vehicles with the antitheft radios are also equipped with on-board computers that monitor lots of interesting things, such as instant miles per gallon, estimated time to destination, average mileage, and distance before an empty gas tank. Unfortunately, the engineers who designed these on-board computers never heard of lithium backup batteries. When the car batteries are disconnected, the computers need to be reset. This can be a very annoying task to perform. If you can avoid this situation, do so. By not disconnecting the vehicle battery, the on-board computer will not need to be reset.

A Word of Caution

If you decide not to disconnect the battery (which I trust I have made a strong case for), be sure to take proper precautions. A car battery is capable of delivering a great deal of current—enough to melt a ring or bracelet (and severely burn the person wearing them) or a tool in seconds. When working on a car, always remove all watches, chains, bracelets, and rings. It is also a good idea to be careful about shorting any "hot" (+12 volt) lines to ground. Since most lines are usually fused, you probably won't damage anything, but it will make one hell of a spark. Be careful!

Passive Restraint Systems

The Mercedes-Benz passive restraint system is worth mentioning if you want to work on the car with the electrical system "hot." This passive restraint system is more commonly known as an "airbag," which inflates during a collision to protect the driver from being thrown into the windshield. While the airbags are designed to be "idiot proof," they are really only "idiot resistant."

It is possible to discharge the airbag by shorting the right (or wrong) lines when the power is active. A discharged airbag will need to be repacked by the Mercedes-Benz dealer, and it is an expensive repair.

DISCONNECTING THE BATTERY

If you decide to disconnect the vehicle battery while performing the installation, the procedure is very simple.

1. Find the right size wrench and loosen the nut that tightens the clamp that holds the power cable to the negative terminal of the vehicle battery. Always remove the negative terminal, since there is no chance of the negative terminal causing a spark if it touches the chassis of the car (because they are both at the same ground potential).
2. After the nut is loose, wiggle the clamp, and it should release.
3. Pull the clamp away from the terminal, and secure it with a cable tie. On rare occasions, the clamp will break. Replacements cost under a dollar, and a few should be kept in stock for such occasions.

A word of caution about working with metal tools around a car battery: While the voltage of a car battery is only 12 volts DC, it is capable of delivering large amounts of current. A car battery can easily melt a screwdriver or wrench in seconds if the tool short circuits the two battery terminals. If the installer is unlucky enough to be holding the tool, he can be burned. By being cautious, this situation can easily be avoided.

CONNECTING THE POWER CABLE

Ground Wire

Most cellular telephones have three wires for power (see figure 6.15). The first is the ground wire, which is usually black but be sure to refer to the manufacturer's installation literature to be certain. The phone can be grounded to the vehicle chassis wherever it is convenient. Be sure to make a good mechanical connection to the vehicle chassis.

1. Using a good, sharp pair of wire cutters, cut the ground cable to the proper length.

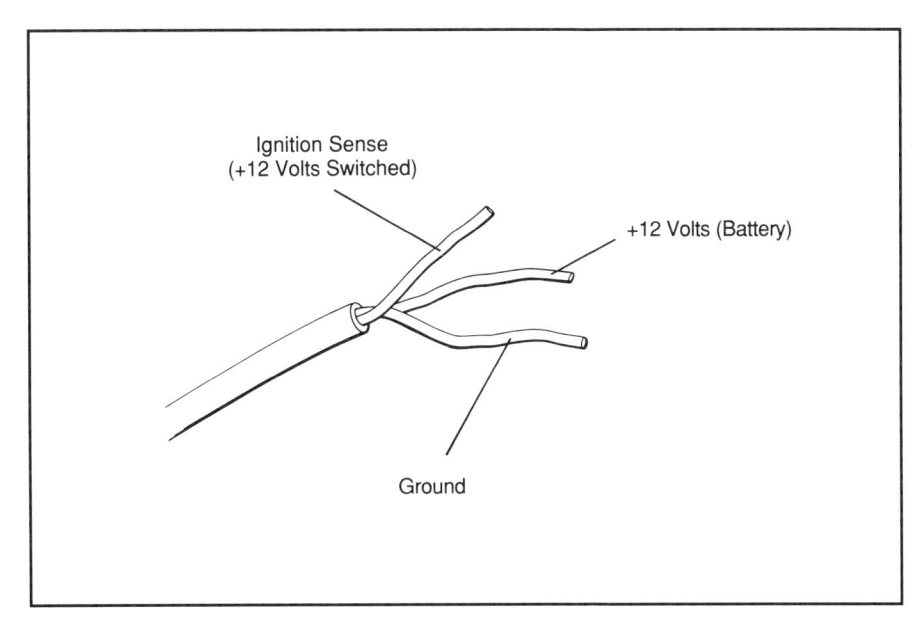

FIGURE 6.15 POWER CABLE FOR A CELLULAR TELEPHONE
The three inner wires are exposed when the insulation is removed.

2. Then remove about 3/8 inch of insulation with your automatic wire strippers (see figure 6.16). Don't use wire clippers to strip the insulation from the wire because they can nick the inner conductor or actually cut some of the strands of wire. This can reduce the current-carrying capability of the ground cable and create problems with the installation.

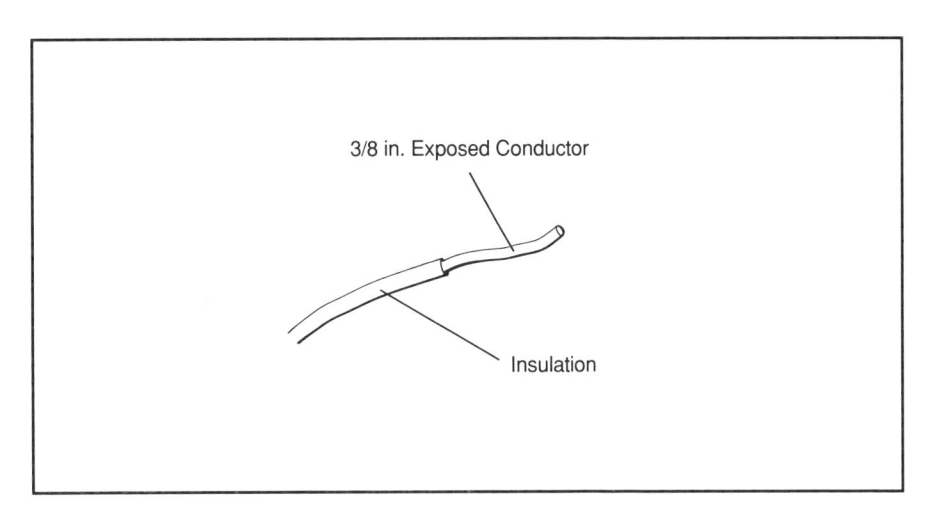

FIGURE 6.16 PROPERLY STRIPPED GROUND WIRE

3. After properly stripping the wire, attach a solderless spade connector to the end of the wire (see figure 6.17). Crimp the connector using the crimper designed to work with that particular connector. Don't use pliers. A properly attached connector should be mechanically secure and should have no exposed conductor.

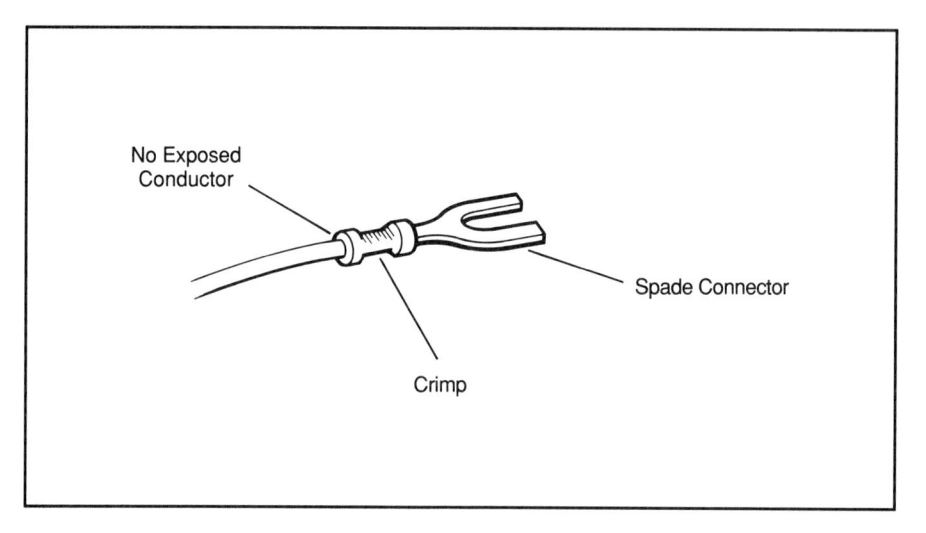

FIGURE 6.17 GROUND WIRE WITH SPADE CONNECTOR ATTACHED

4. If an existing metal bolt is available, remove it and rough up the surrounding area with a piece of emery cloth.
5. Put the bolt through the spade connector and return the bolt to its original location (see figure 6.18).
6. If no such bolt is available, drill a small pilot hole in the vehicle chassis (where it won't be seen, of course) and attach the ground wire to the chassis with a self-tapping screw. A good time-saving trick is to use a battery-powered screwdriver to drive the screw. With one quick push of a button, the screw is in place.
7. Be sure there is no corrosion around the contact point, and use silicon sealer to prevent leaks.

Running the Power and Data Cables

It makes sense to run the power cable and the data cable at the same time. It will not only save installation time, but it is a good idea to run these cables next to each other. This will make

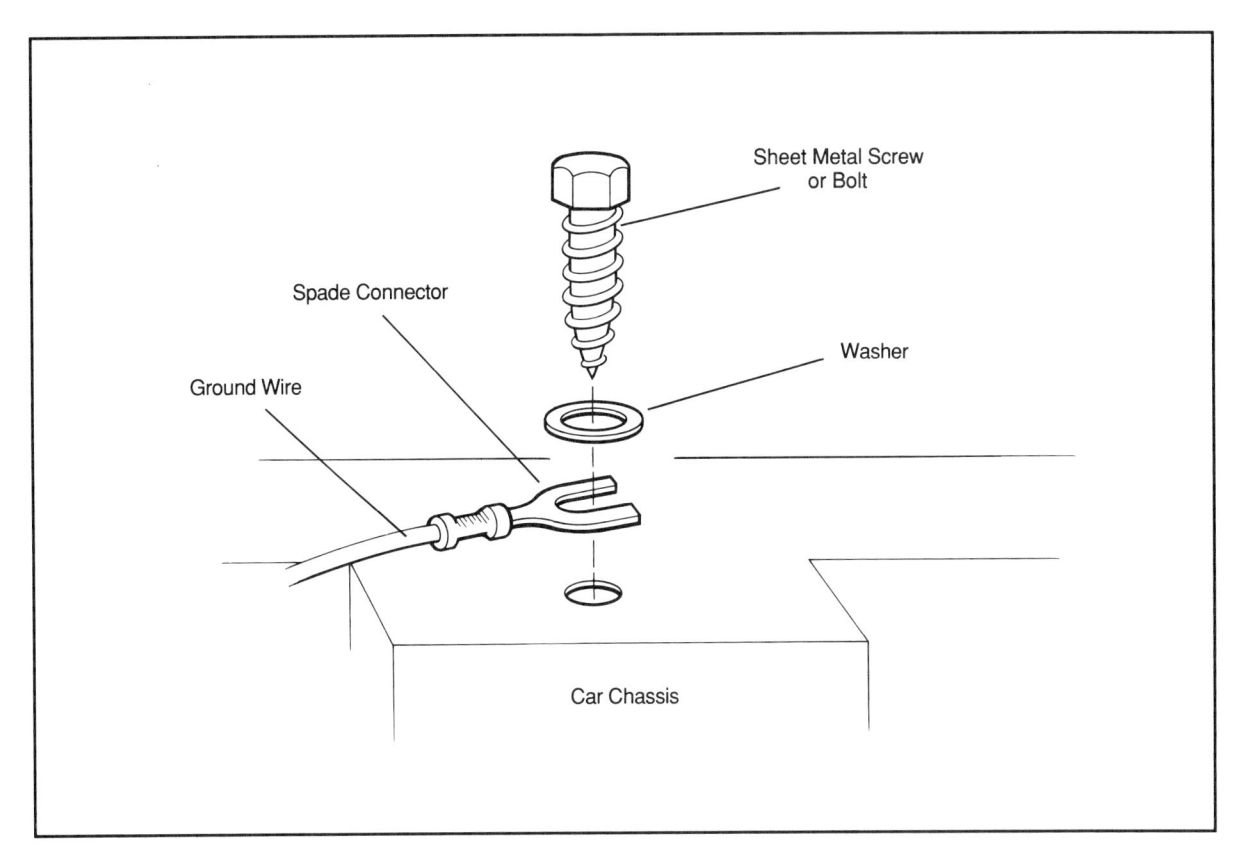

FIGURE 6.18 GROUNDING THE CELLULAR PHONE TO THE CAR CHASSIS

troubleshooting at a later date that much easier. The remaining two power wires and the data cable must be run from the trunk of the car to the interior. Using a piece of electrical tape, attach the power wires and the data cable to a plumber's "snake." Look around the trunk for an opening to the interior of the car that is large enough to push the snake through, and run the power and data cables to the interior of the vehicle.

The door kick panels of most cars can usually be taken off by removing a few screws (see figure 6.19). This allows the installer to run the power and data cables under the carpet or kick panels. Watch out for other vehicle power cables while running the power cable. The installer should be sure that the "snake" doesn't damage any existing vehicle cables.

It is also a good idea to avoid running the power and data cables parallel to any existing vehicle cables. This will minimize the introduction of noise into the cellular phone from the vehicle's

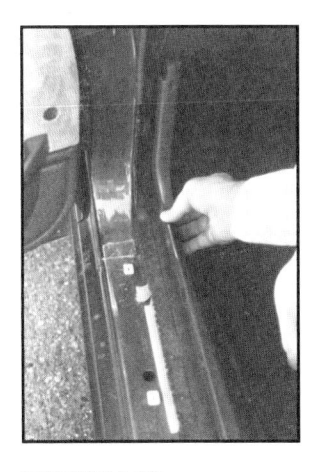

FIGURE 6.19
DOOR KICK PANEL REMOVED

electrical system. Also, many modern "high-end" cars have some of their electrical functions controlled by a microprocessor. In some rare cases, the data flow from the control head to the transceiver can disrupt the vehicle's microprocessor operation. By not running the data cable parallel to any of the vehicle's cables, the problem will almost never exist.

Once the power and data cables are in the interior of the vehicle, separate the data cable from the power cable. Run the data cable to wherever the control head is going to be mounted, and leave it to be connected later. It is very important to be careful with the data (and power) cables when pushing them through small holes and under carpets. One of the more common causes of nonworking installations is a damaged data or power cable. Be sure that the cable has not been cut by any sharp metal edges in the car and that the connectors are not "stressed" too much. Most connectors can be pulled off the cable without too much effort. If the cable is damaged, it is usually a good idea to replace it. Don't try to repair it! This will usually cause more trouble than it is worth. If the radio is damaged by this kind of repair attempt, the manufacturer usually won't cover the repair under the warranty.

Ignition Sense Wire

The second line we will connect is the ignition sense wire. (See the manufacturer's installation manual for the correct color code.) Since this wire turns the cellular telephone on and off with the vehicle's ignition, you need to locate a 12-volt source that is turned on and off with the vehicle ignition. The best place on most cars is the car radio, which provides a "clean" (interference-free), simple power connection. Most car radios are set up like cellular phones. They have two 12-volt inputs: one that is always active (or "hot") and one that switches the vehicle's 12 volts when the car's ignition is turned on and off. Find the switched line with a test light. To do this quickly, find the car radio switched power line. It should lead to the fuse block and be clearly marked. To be sure that you have the right power line, attach (with the alligator clip) the wire from the test light to a bolt in the car that is grounded. Using the sharp point of the test light, pierce a small hole in the "suspected" power line. When the car's ignition is on, the light should glow, because there is 12 volts present on the line. When the ignition is switched to off, the test light should go out.

There are a few exceptions to this (of course). Some of the newer, very high-end cars have microprocessors that operate some of the vehicle's accessories. The accessories are controlled by data buses. The new Jaguar is one example of a car that uses this approach. Be sure that you don't make the mistake of thinking this is a 12-volt line. These data buses operate at 5 volts and, if they are shorted, can destroy the vehicle's computer system. In some vehicles, if the computer is destroyed, the car won't start, the windows won't work, and the installer will have a large bill to pay. But don't lose faith in the installation business. Jaguar (and the rest of the manufacturers who use these data systems) have a 12-volt line that is designed specifically for use with cellular telephones. It is usually well marked and described in the operator's manual. Sometimes there is even a spot in the fuse block for a cellular phone.

The first time you perform an installation on a particular vehicle, it is a good idea to be sure that the phone has not overloaded the vehicle's electrical system. Although this is very unlikely, be sure that no fuses are blown when a conversation is taking place. (The phone consumes the most current during this phase.)

After the switched 12-volt line has been properly identified, follow these steps:

1. Remove the ignition sense wire from the "snake" and cut it to the proper length. Do not strip any insulation from the end of the wire.

2. Instead of splicing into the car's radio power line, we will use a special connector for this purpose. This connector is called a *3M connector* and is available though your electronics "jobber." This special connector "snaps" over the end of the ignition sense wire.

3. The connector then clamps onto an existing wire and provides an electrical connection between the two (see figure 6.20).

    This entire process takes about 15 seconds and provides a good, reliable electrical connection. I don't recommend soldering these wires, because the soldered connection is no better than the one made by this special connector. Soldering is time-consuming, and there is always the risk that the soldering iron will damage a part of the vehicle's interior.

4. Be sure to attach the ignition sense wire to an existing cable harness with cable ties. This keeps the cable out of sight and ensures that the cable won't be pulled and disconnected.

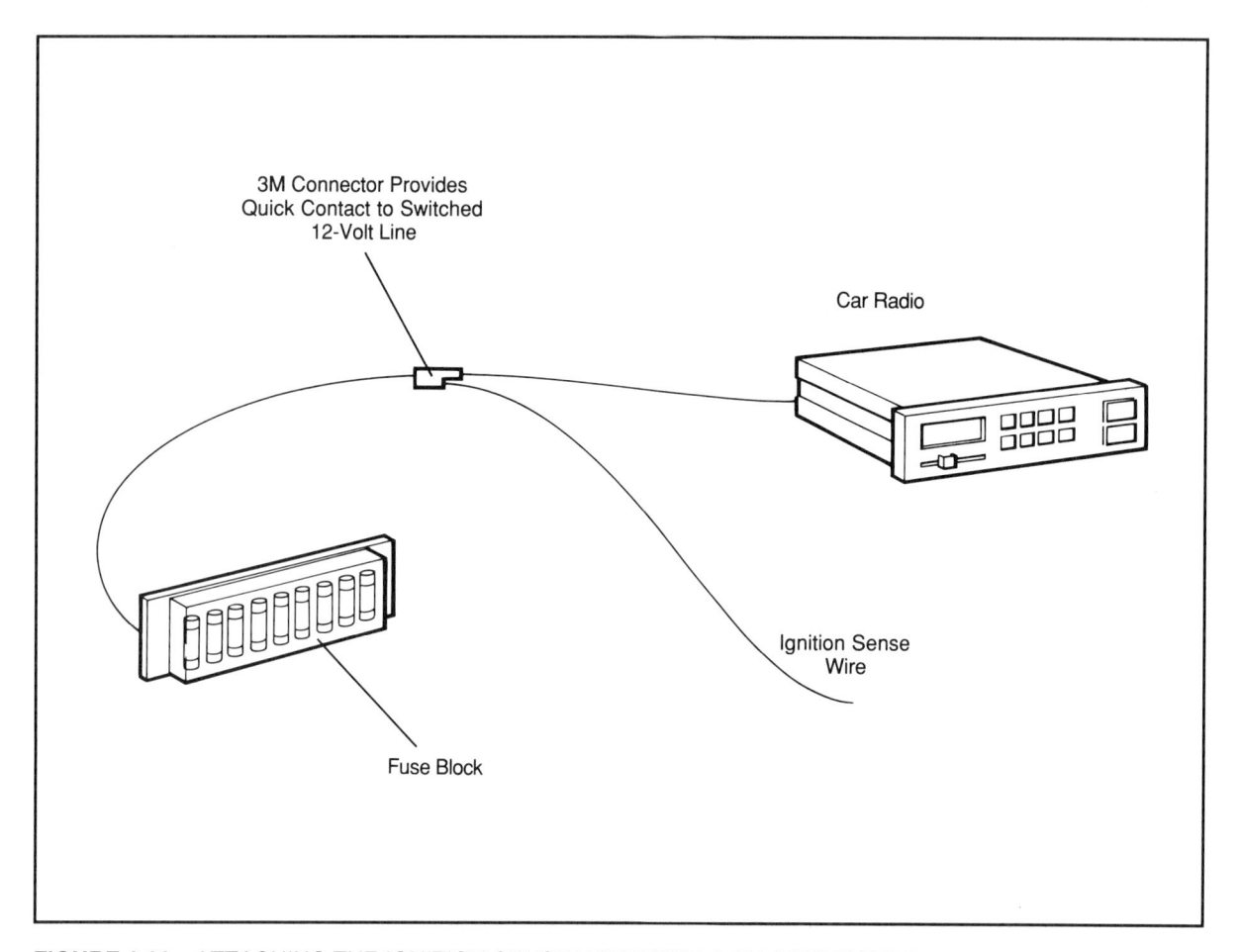

FIGURE 6.20 ATTACHING THE IGNITION SENSE WIRE WITH A 3M CONNECTOR
The car radio's switched power line is the "best" place to attach the phone's ignition sense wire.

Constant Power Wire

The last wire to be connected is the constant power. The color code for this wire varies from manufacturer to manufacturer. It is important to refer to the installation manual to be sure that you have the right line. Applying voltage to the wrong wire can seriously damage the cellular phone. This power line consumes a small amount of power even when the phone is turned off. Don't worry about draining the car battery, because the phone consumes less power than the car's clock. The battery will discharge itself before the phone will drain it. This power is used to keep the phone's repertory memory active. If the phone is left without power for a long period of time, it will "forget" the numbers that have been stored in memory.

To connect the constant power wire, follow these steps:

1. The first thing the installer must do is locate a constant power source. It is possible to find a place under the dashboard that will supply a constant 12 volts, but I don't recommend this practice. There is a lot of stray RF energy under the dashboard of a car that can cause interference and create noise when the cellular phone is used.

    The best place for a good source of constant 12 volts is the battery. It requires a little more work on the part of the installer, but it will produce a cleaner installation. By obtaining a constant power source from the battery, a minimum amount of noise will be introduced into the cellular telephone.

2. Attach the remaining wire to the "snake" with electrical tape and look for a hole in the vehicle's firewall (the wall between the engine compartment and the passenger compartment). In some vehicles, there are rubber plugs that seal holes in the firewall.

3. Remove the plug, and place a grommet in the hole. This will prevent the cable from being damaged by any sharp metal edges.

4. After installing the grommet, push the "snake" through the hole.

5. Remove the power wire from the "snake," and run it to the car battery.

6. Connect the B+ or constant power wire directly to the vehicle's battery. Prepare the wire in the same manner the ground wire was prepared.

7. Attach a spade connector to the end of the wire.

8. Loosen the nut that holds the heavy power cable to the car battery terminal.

9. Put the spade connector behind the nut and tighten it (see figure 6.21). This will provide a secure mechanical connection to the car battery.

10. Be sure to attach the cellular phone's power wire to another cable harness in the engine compartment with cable ties. This will keep the wire from being caught in any of the engine's moving parts.

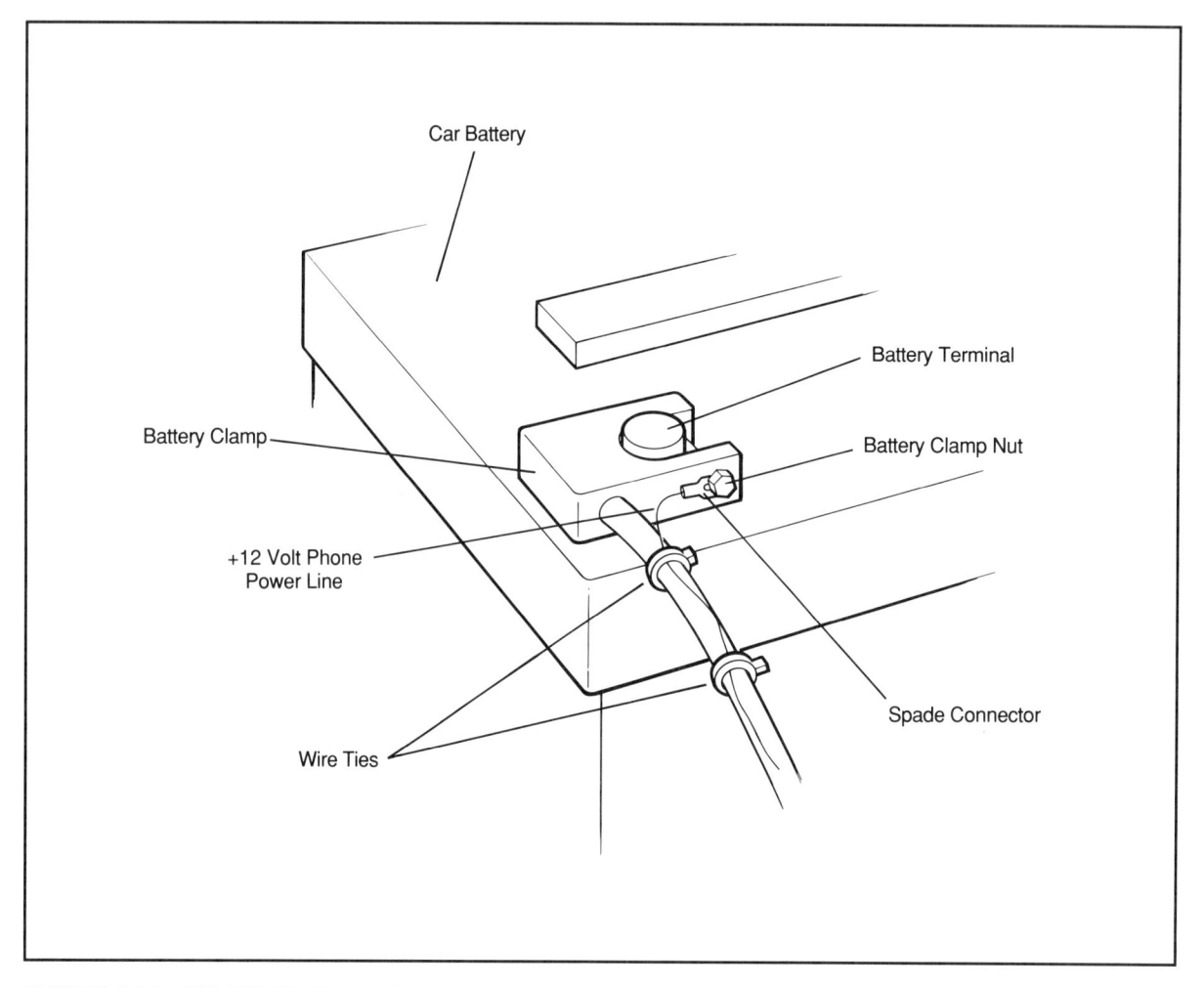

FIGURE 6.21 CELLULAR PHONE'S 12-VOLT POWER LINE CONNECTED TO CAR'S BATTERY TERMINAL

INSTALLING THE ANTENNA

Proper installation of a cellular antenna is one of the most important parts of the job. The antenna is the means by which a cellular phone communicates with a cell site. The antenna installation can make the difference between a cellular phone that works and one that doesn't.

Glass-Mount Antennas

Mounting the Antenna and Coupling Box. Once a location has been selected, the area where the coupling box is to be mounted must be completely cleaned. Windex or a similar glass cleaner should be used to remove dirt and oil from the window.

Most manufacturers of glass-mount antennas include a small pad soaked with alcohol. This pad is used to clean the window a second time. Different manufacturers use different materials for mounting their antennas and coupling boxes to the glass surface. Some use a silicon-type adhesive, while others use a double-sided tape. The double-sided tape allows for a cleaner and quicker installation than the silicon-type adhesive. However, once the antenna is attached, it is there to stay. It must be removed with a razor blade and cannot be remounted. **Remember:** When using double-sided tape, an installer does not get a second chance to mount the antenna.

Silicon adhesive takes some time to cure. Read the manufacturer's instructions for exact times. If the installation area is too cold, the adhesive will not cure properly, and the antenna could fall off sometime in the future. A good rule of thumb is to keep the installation area at a temperature that is comfortable to the installer. This adhesive (or tape) will have to hold the customer's antenna on for a long time, so take proper care and do it right.

Running the Coaxial Cable. Proper routing of the coaxial cable can be the difference between a first class installation and an angry customer. The cable can usually be hidden by routing it through the vehicle headliner and down one of the vehicle's side moldings. Extra care should be taken to carefully pull down the headliner and firmly attach the cable to the metal roof. Duct tape will secure the cable properly. If the cable is simply laid on the headliner, it may vibrate when the vehicle is moving. This is a great way to make a customer angry and to destroy your credibility as an installer. After carefully reattaching the headliner, look for a way to route the cable behind the side molding by the rear window. Then route the cable to the vehicle trunk by sliding it behind the back seat.

Cable clips, which are sold by a number of manufacturers, are small plastic cable holders with a piece of adhesive mounted on a flat portion of the holder. They allow an installer to secure a cable so that it will not be visible. I recommend the use of these cable holders to properly route antenna cable from the coupling box to the vehicle's side moldings if access to the vehicle headliner is not practical.

While there are varying opinions on the matter of cable length, I suggest trimming any excess cable before mounting the coaxial connector. While the signal loss in good coaxial cable is small, removing extra cable can only help improve the power radiated by the vehicle's antenna.

Attaching the Coaxial Connector. Now comes the final and most critical part of installing any vehicular antenna—attaching the coaxial connector. The first step is to buy a crimping tool. No installation facility should be without one of these devices. Under no circumstances should you attempt to crimp a connector with a pair of pliers. Most antenna manufacturers supply a solderless-type connector. When installed properly, these connectors work very well. When installed improperly, they don't work at all.

1. First, trim the cable to the dimensions recommended by the manufacturer (see figure 6.22 for one example). Be sure not to nick the shielding with the stripping tool. Do not use "dykes" or wire cutters. A high-quality stripping tool is an absolute requirement.

2. Slip the crimp sleeve over the cable. If you forget to perform this step, you will have to remove the connector and attach a new one.

3. Place the center conductor into the center contact of the connector. Be sure the center contact is flush against the dielectric material. If the center conductor is not the right length, the center contact will not fit properly and a bad connection will result.

4. Crimp the center contact with the proper crimping tool. **Remember:** Use the proper crimping tool, not pliers.

5. Flair the coaxial conductor's shielding and push the connector housing over the center conductor. Be sure none of the small shielding wires touches the center conductor. This will short the transceiver and possibly damage it.

6. Slide the crimp sleeve (the one that was pushed up the cable in step 2) so that it touches the connector housing, and crimp with your tool.

7. Trim the excess shielding from the edge of the crimped connector. Be sure to pick up the trimmings from the floor of the vehicle. It is this attention to detail that generates referrals.

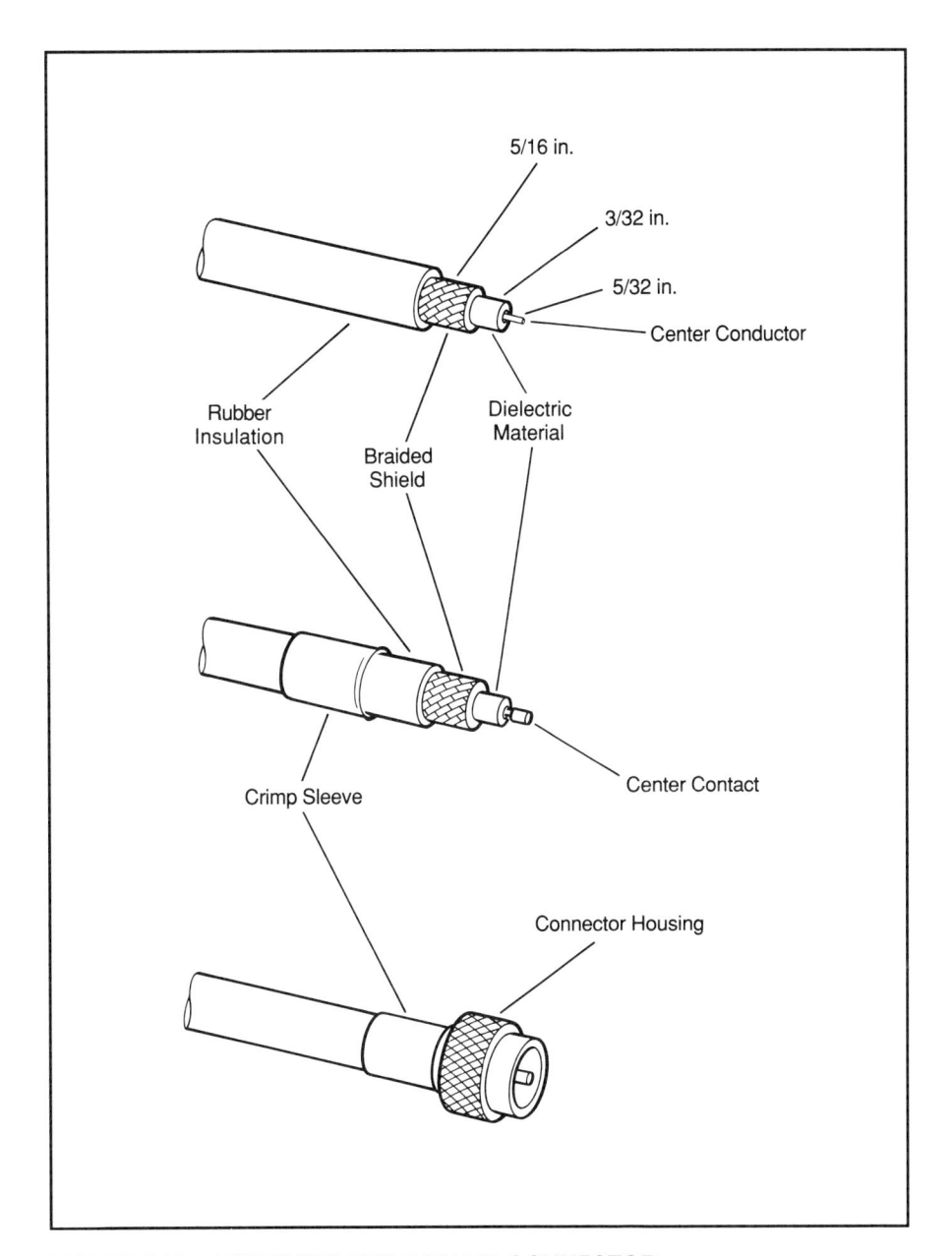

FIGURE 6.22 ATTACHING THE COAXIAL CONNECTOR
A properly prepared cable makes attaching the coaxial connector a simple task.

These directions are somewhat generalized, but they should give you a good idea of what is involved in attaching a connector. With a bit of practice, a good installer can do it in about two minutes.

Checking the Installation. Take a few minutes to check the installation. Are there any "kinks" or sharp bends in the coaxial cable? Is a minimum of cable showing? Are there any scraps from the cable still in the vehicle? Is the vehicle headliner properly reinstalled? Would you be satisfied with this installation if this were your car? *Remember:* Attention to detail is the difference between a fair installation and a professional one.

Elevated-Feed Antennas

Drilling the Hole. The first time you drill a hole in a car, your hands will probably shake a bit. If care is taken, however, the process is quite painless. First, place a fender cover over the fender. The last thing any installer wants to do is scratch the vehicle's paint. I can say from experience that repairing a scratch on a Rolls-Royce will put a serious dent in an installer's paycheck. With the fender cover in place, select a place to drill the hole. Be sure there is enough flat area with which to work. There should be no curve or ridges in the metal where you plan to mount the antenna. If no such surface is available, suggest to the customer that he/she select either a glass-mount or roof-top antenna. Once the spot is found, measure to find the center of the fender so that the antenna will look good after it is mounted. Don't guess, because you will usually end up being wrong and there is no chance to "try again" if the hole is in the wrong place.

After the spot is marked with a grease pencil, use your spring-loaded scoring tool to put a small dent in the spot. This dent will provide your drill bit with an indentation in which to sit until the bit "digs" into the metal. I have seen installers who just mark the spot and attempt to drill. Since the bit has no dent in which to sit, it can "wander" from the proper place. In the best case, the hole is in the wrong place. In the worst case, the drill "dances" across the fender or trunk of the vehicle and someone (usually the installer) ends up paying for an expensive paint job.

Before drilling any holes, put on a pair of approved safety glasses. When drilling metal, small burrs are ejected from the drill bit in all directions. Getting one of these burrs in your eye can blind, so take the time to protect yourself.

With the spot marked and properly "dented," drill a small pilot hole with a very sharp drill bit. Apply firm, even pressure until the bit is through. After drilling the hole, use a shop vacuum or a portable unit to pick up the metal shavings before they can damage the vehicle or installer.

After the pilot hole is completed, use a hole punch to achieve the proper hole diameter. Most elevated-feed antennas require a fairly large diameter hole (up to 15/16 inch). While drill bits are available in these large diameters, I prefer to use a hole punch. Drill bits are always difficult to center in exactly the right place. The installer also runs the risk of the drill bit "jumping" out of the pilot hole and damaging the vehicle's paint. An installer can position the hole punch in the desired location and slowly punch the hole. It is also very difficult to make a mistake with these tools. After punching the hole, remove the burrs with a soft paint brush or vacuum. It is again important for the installer to protect his/her eyes while doing this. Using a file, remove some of the paint from the mounting surface under the hole. This will ensure a good electrical ground for the antenna.

Mounting the Antenna and Running the Cable. Mount the antenna according to the manufacturer's instructions. Next, run the cable behind the carpet to the transceiver. Do not attach the connector until the cable has been properly routed. Don't be afraid to use cable ties to secure the cable while running it. Cut the cable to length after running it, and attach the proper connector. (See pp. 126–27 for instructions.)

Roof-Top Antennas

Roof-top antennas are the most difficult to install. They are also the most time-consuming (and need to be priced accordingly). After careful planning, the job begins.

Drilling the Hole. Let's assume that mounting over the dome light is acceptable to both the installer and the customer. First, remove the plastic cover on the dome light; it should snap off. Inside, there are usually two screws that hold the assembly to the roof. Remove them, and the assembly should drop off. If the wires are attached with plugs, remove them, and put the light assembly on the workbench. With a measuring tape, find the center of the roof. Don't guess or "eyeball" it. Holes in the wrong part of the roof are expensive to fix. After the center has been marked, use the spring-loaded scoring tool to make a spot for the drill bit to start. The process of drilling the hole is a two-person job: a driller and a spotter are required. The driller, of course, drills the hole. The spotter sits in the car and watches to be sure the drill bit does not appear in the wrong place and make a hole in the vehicle headliner (which is expensive to replace). It is a good idea to put a thin piece of metal or wood between the roof and the headliner while the hole is being drilled. Once the pilot hole is drilled, a hole punch is used to make the hole the correct size for the antenna. Be sure to read the manufacturer's instructions, because some antennas require different size holes.

Mounting the Antenna and Running the Cable. Since each manufacturer uses a different method for mounting its antennas, detailed instructions won't be included here. Just be sure to follow the manufacturer's instructions when mounting the antenna.

The coaxial cable can usually be "snaked" over to one side of the car where the headliner can be slightly retracted from the roof. Be sure to secure the cable so it doesn't vibrate when the vehicle is moving. The installer can stick his/her hand between the headliner and the roof and grab the cable. Remember not to attach the coaxial connector until the cable has been run back to the transceiver. The best way to do this is up to the installer. One of the better methods is to remove a piece of side molding and run the cable to the floor. It can then be placed under the carpet and routed to the trunk from behind the rear seat. If the transceiver is mounted under the front seat, the cable is that much easier to run. After this process is complete, attach the coaxial connector according to the instructions on pp. 126–27.

TESTING AND PROGRAMMING THE EQUIPMENT

Before mounting any of the equipment, it should be tested and programmed. A good way to test the equipment is to remove all of the cables (with the exception of the power cable that we have connected) and hardware and connect them together as the manufacturer specifies in the installation manual. Connect the control head to the transceiver with the data cable. Before connecting a power cable to the transceiver, determine whether the cellular unit is programmed through the keypad (or service handset) or a PROM. If a PROM is required, use your NAM programmer to input the proper information into the PROM. See Chapter 3, "Test Equipment and Tools," for help. Install the PROM into the transceiver, following the manufacturer's instructions.

If the unit is programmed through the keypad, attach the power cable to the phone and turn it on. Manufacturers all have their own power connectors. It is a good idea to have one for each piece of equipment that you test and to keep them at the service bench. This will speed up the programming time, since you won't have to search for cables every time you want to program or test a phone. Follow the manufacturer's programming instructions, and input the necessary data into the phone.

Auto Test

After programming the unit, it is a good idea to test it before installing the equipment in the vehicle. There are several ways of doing this. One of the best ways is to attach the unit to a cellular test center through the antenna connector and then to run the auto test program (see figure 6.23). The program will test all of the

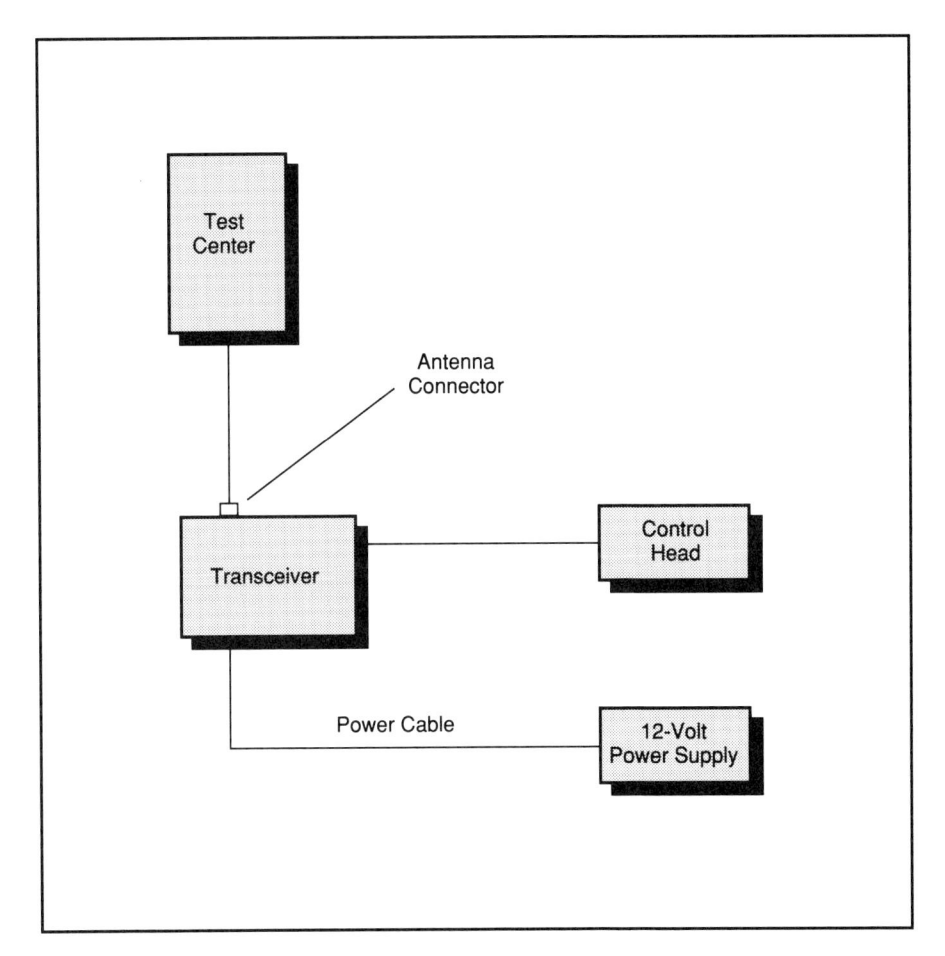

FIGURE 6.23 BLOCK DIAGRAM OF A CELLULAR TELEPHONE CONNECTED TO A TEST CENTER

phone's functions and return an alarm if any of the specifications are out of tolerance. If any of the specifications are not met, either adjust them using the manufacturer's service manual for reference or return the equipment to the manufacturer.

Before returning the equipment to the manufacturer, it is a good idea to call their service department. By describing the problem to them, they may be able to give you a "quick fix" and save you the effort and expense of returning the equipment. If you plan to return the equipment to the manufacturer, be sure to clearly detail the problem so that their service people can quickly determine the cause of the problem and return the repaired equipment to you.

Manual Handoff Test

If the cellular phone passes the auto test, try a few manual handoffs. The test equipment manual will tell you how to do this. Set the RSSI level to about −100 dbm and attempt a handoff. The phone should have no trouble at this signal level. Next, reduce the signal level to −110 dbm, and repeat the handoff. Further reduce the signal strength to −114 dbm, and repeat the test.

The phone's ability to perform successful handoffs at low signal levels indicates the unit's ability to hold and process calls in areas with low signal levels. This is very important to the proper operation of the phone. If the phone fails any of these tests, it should be returned to the manufacturer with a clear description of the defect.

If no test equipment is available, there is a way to determine if the phone is operational. Using the A/B switch, select the alternate carrier (the carrier that the customer will not be registered to use), and place a call. Because the customer is not registered with that carrier, the call will be routed to a special announcement that lets the customer know that he/she is not a valid subscriber. In order to hear the announcement, the phone must be capable of processing a call properly. I don't recommend this test in place of owning the proper test equipment, but if you are in the field or your test center is out of order, it is better than nothing.

MOUNTING THE TRANSCEIVER

Next, the transceiver should be mounted. As discussed earlier in the chapter, several options exist when selecting a transceiver location. The most common is in the vehicle trunk. This is my favorite location for an installation because it is one of the easiest to perform and provides the most flexibility.

All transceivers are supplied with a mounting plate. With the mounting plate in hand, look around the trunk for possible locations. This step will become unnecessary after you become familiar with the different models of vehicles. The best location for the transceiver is in one of the corners of the trunk. By placing the transceiver in the corner, it is kept out of the customer's way and doesn't preclude the use of the trunk from its original purpose. In figure 6.24 we see a typical installation location for the transceiver. Note that the cables are kept out of the way so that they won't be damaged by luggage or other articles rolling around in the trunk.

FIGURE 6.24 TYPICAL INSTALLATION LOCATION FOR A TRANSCEIVER

While you are looking for a transceiver installation location, take note of the location of any of the vehicle's cabling and the gas tank. There is nothing quite as embarrassing as drilling a hole or putting a screw into a customer's gas tank. However, if you do enough installations, you are eventually going to drill a hole in a gas tank. While it is not dangerous, it is very time-consuming to repair. When this happens, the first thing you should do is to let the customer know "up front" what has happened. If customers find out about the damage at a later date, they will be angry and may even bring legal action. Admitting the error "up front" will usually diffuse the customer's anger. We will discuss gas tank repair in Chapter 9, "Troubleshooting and Repairing Installations."

Once a location has been found, follow these steps to mount the transceiver:

1. Mark the locations where the mounting screws will go.

2. Remove the mounting plate and make pilot holes for the screws. This can be done with either a battery-powered drill or a hole punch and a hammer. Before doing this, check once again for the gas tank and other cables. If you are going to drill, don't forget to wear safety glasses.

3. After drilling the pilot holes, vacuum the metal filings so you won't get cut.

4. Next, place the transceiver mounting plate in the proper location, and attach it with the screws supplied by the manufacturer. This is another job for the battery-powered screwdriver. Attaching the transceiver mounting plate with screws is the recommended method for trunk and rear-deck (in hatchback vehicles) installations (see figure 6.25). A good mounting shortcut is to use self-tapping screws (available at hardware stores and from parts jobbers). However, be particularly careful when using self-tapping screws. They don't give you a very good "feel" for what you are drilling into. Some manufacturers supply these screws with their mounting hardware. These screws have small cutting blades on their heads that drill the pilot hole. By using a battery-powered screwdriver, the mounting time for a transceiver plate is cut down to less than a minute. The procedure is the same for both transportable and mobile-only transceivers.

If the customer wants the transceiver mounted under the seat, it is not usually practical to mount the transceiver plate with screws. Instead, velcro can be used to securely fasten the transceiver to the vehicle floor, as shown in figure 6.26. Using "hot glue" or a similar adhesive, fasten one side of the velcro to the vehicle's carpet. Attach the other side to the transceiver itself. Don't spare

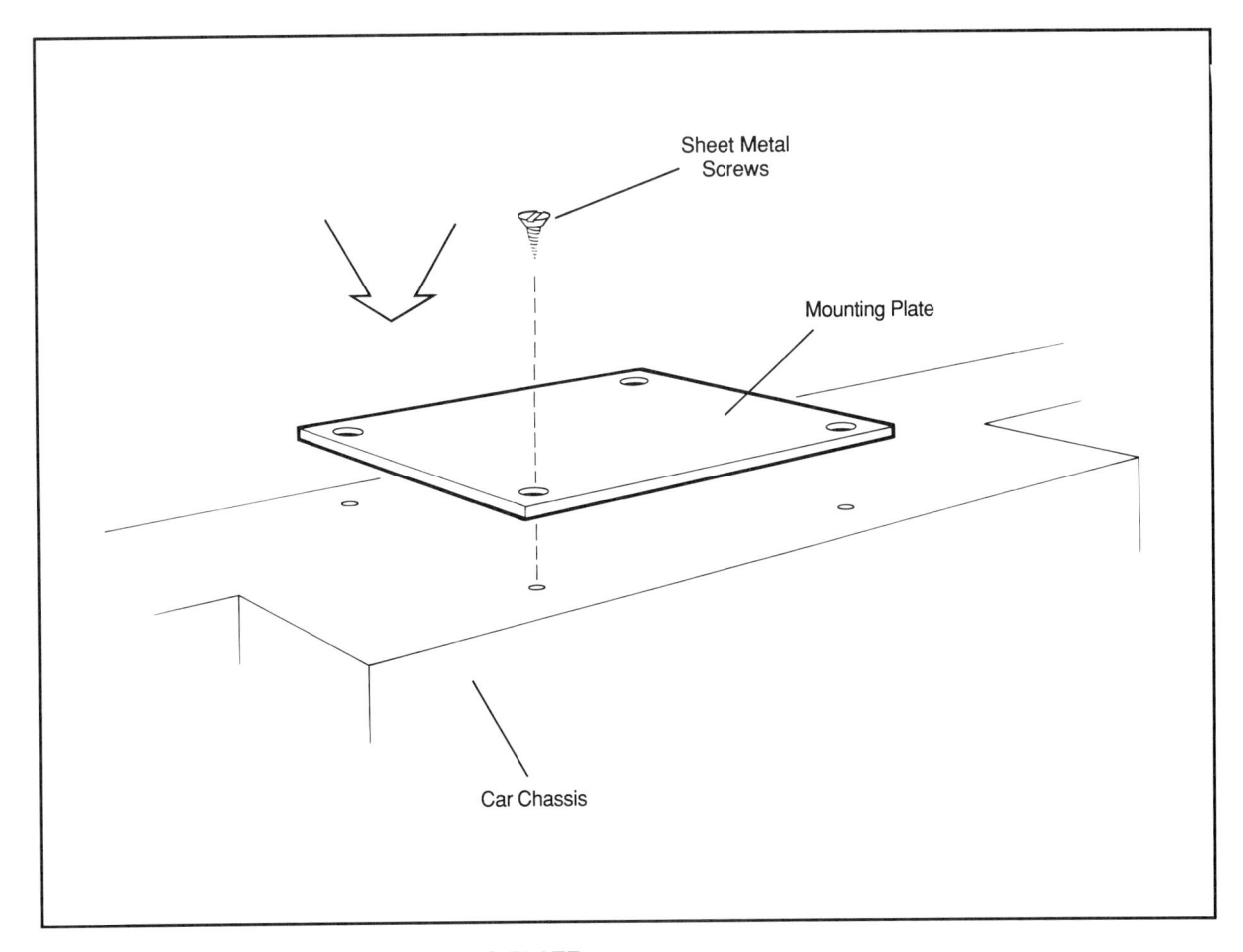

FIGURE 6.25 TRANSCEIVER MOUNTING PLATE
The mounting plate is attached to the car chassis with sheet metal screws.

the velcro. Slide the transceiver under the seat and "mate" the two pieces of velcro. The transceiver will be held securely in place by the velcro.

MOUNTING THE CONTROL HEAD

The control head is the part of the installation that the customer will see most frequently. It is therefore important that extra care is taken during this part of the installation. One of the most important aids that you will use during the installation of a control head is velcro. Velcro is a great mounting aid that eliminates the need to drill a hole in the customer's car. Holes in

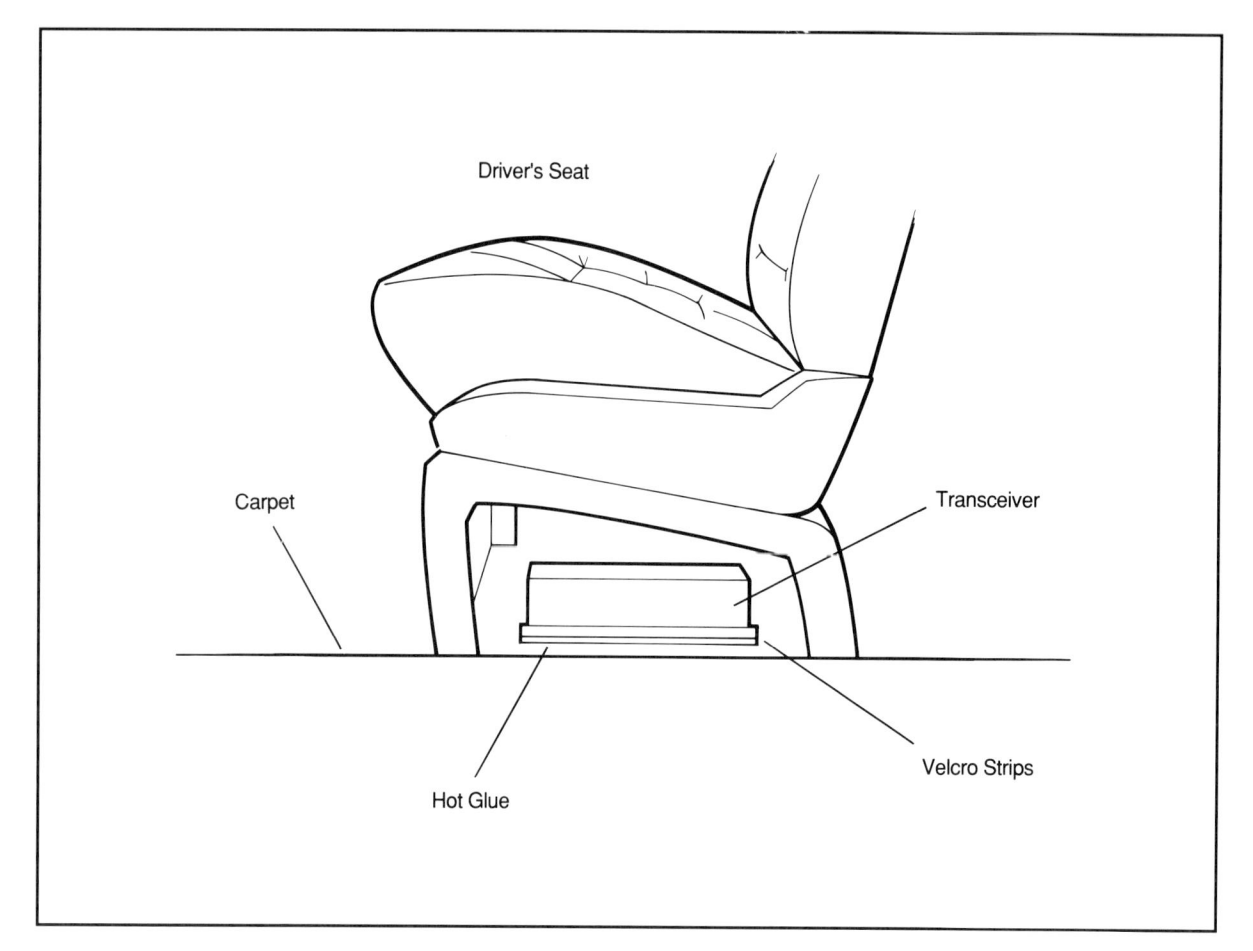

FIGURE 6.26 TRANSCEIVER MOUNTED UNDER SEAT WITH VELCRO

the interior of the car are much more difficult to repair than holes in the car body. There are several products available to repair holes in vinyl or on dashboards, but they never look as good as new.

As I stated earlier in the chapter, my favorite place for a control head to be mounted is between the front seats on the center console. Most vehicles have some kind of center console with a flat area that is suitable for mounting the phone. This center console offers a number of advantages for both the installer and the customer. If the phone is placed in this location, everyone in the vehicle has good access to it. By finding a flat location on the center console, the need to use a mounting bracket or to drill a hole is eliminated. Follow these steps in mounting the control head:

1. Clean the mounting area with alcohol to remove any grease in the area. The adhesive on the back (sticky side) of the velcro will not stick to anything with Armor All™ or similar products on it. The Armor All™ must first be removed with a solvent before the velcro can be applied.
2. Carefully mark the intended mounting area, and apply a few strips of velcro.
3. Place a few strips of the other side of the velcro on the control head base. It is a good idea to cover the entire mounting surface of the base with velcro. This will fasten the head more securely than just one strip.
4. Now place the control head in the proper location. Check to be sure that there is no velcro visible to the customer when the head is mounted.

With a velcro mounting, the control head can be removed easily and stored in the trunk or tucked under the seat when the car is left unattended.

When installing the control head in an American sedan, the best location is usually on the floor on the center "hump." Some manufacturers supply their equipment with mounting hardware. Others choose to let the installation facility provide its own. There are several good manufacturers of mounting hardware listed in Appendix B, "Directory of Manufacturers." Use a mounting bracket that elevates the control head a few inches off the floor so that it is easier for the driver to reach (see figure 6.27). Take some time to chose a good location for the mounting bracket. Here are a few questions to keep in mind while looking for a good location:

1. Can the driver easily reach the phone?
2. Does the mounting bracket block the seat, ashtrays, stereo, or other accessories?
3. Will the bracket be in the way of passengers?
4. Does it look good?
5. Can the driver easily see the phone while driving the car?

After satisfying these questions, mark the location of the mounting screws. Using a hole punch, perforate the floor of the

FIGURE 6.27
CONTROL HEAD MOUNTED ON LONG MOUNTING BRACKET

car so that the mounting screws have pilot holes. Be sure to use large enough screws to securely hold the mounting bracket to the floor. Remember that when the car moves, the vibrations of the car will be amplified with the long mount. Feel free to use self-tapping screws to save time. Don't worry about leaving holes after the control head is removed; the carpet will do a good job of hiding them.

The last common location for a control head is the dashboard. This is my least favorite location because it poses the most risk. Most modern cars have a myriad of electronics located behind the dashboard. Drilling into a cable harness or an electronic module is expensive and time-consuming. If you must mount the control head on the dashboard, use velcro instead of drilling holes. Be sure not to obstruct the view or access to any equipment.

INSTALLING THE HANDS-FREE MICROPHONE

Selecting a Location

The hands-free microphone installation is very critical to the proper operation of a hands-free system. The microphone picks up all of the sound in the vehicle, in addition to the voice of the driver. Since this will significantly affect the overall sound quality of the hands-free system, placement is very important. The most common location is on the sun visor (see figure 6.28). This location is ideal because it is directly in front of the driver and the cable is easy to run.

FIGURE 6.28 HANDS-FREE MICROPHONE MOUNTED ON SUN VISOR

If the customer objects to the sun visor location, an alternate location is the roof side molding (see figure 6.29). This location still makes it easy to run the cable, but the microphone is not directly in front of the driver. Finally, the steering column is also a good place to mount the hands-free microphone. It is out of the way and is in fairly close proximity to the driver's head.

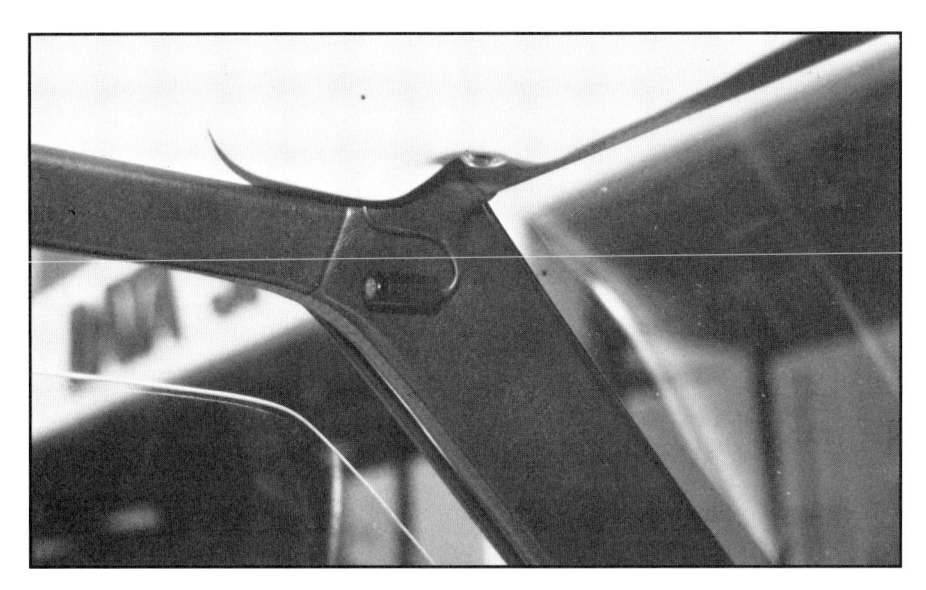

FIGURE 6.29 HANDS-FREE MICROPHONE MOUNTED ON SIDE MOLDING

Convertibles offer some special challenges. The visor cannot be used for the installation because of the noise from the wind. A good alternate location is somewhere on the dashboard. Be sure to stay away from air conditioning ducts, speakers, and other sources of noise. The dashboard location works out fairly well because the microphone is still in front of the driver, and it is kept out of the wind. You should be sure the customer is aware that the hands-free system will not work very well when the roof is down or if there is a great deal of ambient noise in the car.

Running the Cable

Installing the hands-free microphone is very easy to do. Simply remove the side molding above the door on the driver's side (see figure 6.30). Attach the microphone to the sun visor with the clip supplied by the manufacturer or with a small piece of velcro. If you plan to use velcro, be sure to clean the visor with a solvent to remove grease and Armor All™. Leaving enough cable so that the

FIGURE 6.30 SIDE MOLDING REMOVED

visor can be moved, run the cable down the side panel and secure it with electrical tape. Replace the side panel, and run the cable to the control head. Most manufacturers have their hands-free microphone inputs on the control head. Be sure to read the manufacturer's instructions before running the cable, because there are a few exceptions. Secure the microphone cable under the dashboard with wire ties so that it is not visible to anyone in the car.

Calibrating the Hands-Free Microphone

The final step in installing the hands-free microphone is calibration. It is also the most overlooked step. A hands-free system is basically the same concept as an office speakerphone and, as such, is a system based on a number of compromises. Even when optimally adjusted, it still sounds artificial and hollow. When improperly adjusted, it is unusable. Be sure that you are familiar with the manufacturer's instructions and recommendations about their hands-free system. Follow the calibration instructions, and if you have questions, don't be afraid to call the manufacturer's customer service people for help.

CONNECTING THE DATA CABLE

We have completed all of the difficult parts of the installation. Now, you simply connect the data cable to the control head and transceiver and do a quick visual inspection of the installation. Data cable configurations vary among different manufacturers, so be sure to refer to the instructions for each individual configuration. Also, be sure that there are no crimped or cut antenna or data cables. After (and only after) you are satisfied that all the cables are undamaged, attach the power cable and turn the phone on.

Turn the vehicle ignition to accessory, and the phone should turn on. The only exceptions to this are phones that have been directly wired to the vehicle battery (hot wired) at the customer's request or transportable phones that have their own internal battery. These types of installations are turned on and off by the power switch on the control head, regardless of the status of the vehicle ignition switch.

MAKING A TEST CALL

After turning the phone on, place a test call to be sure that the phone works. Many carriers have a 1kHz test tone that can be called to test a phone's operation. Check with your local service provider for details.

Test both the hands-free system (if the phone is equipped with it) and the handset. While the call is being placed, have someone else "jiggle" the antenna and data cables. Remember that "jiggle" does not mean "yank." Small, light movements with the cable are sufficient. You don't want to introduce any new problems. If there are any loose connections, they will create some static and alert you to potential problems before the customer leaves the shop.

CHECKING OUT THE CUSTOMER

Once you are satisfied that the installation is a good one, it is time to "check out" the customer. This means that you should give the customer some basic instruction on the operation of the phone. My philosophy is that the checkout should be kept simple. It is not necessary to show the customer how to use the advanced features of the phone (unless they ask, of course). The process is too long, and they will forget most of what you tell them anyway. Keep the instructions basic. Here is a suggested list of items that you should explain to the customer:

1. How to turn the phone on and off
2. How to place and receive a call
3. How to store and retrieve a number from memory
4. How to adjust speaker, earpiece, and ring volumes
5. How to remove the control head

If customers have any questions, they can certainly call later. You should give the customer your card and encourage continued contact.

REVIEW OF INSTALLATION PROCEDURES

Let's review the steps for installing a cellular phone:

1. Be sure to check all the electrical and mechanical functions of the car before beginning the installation. This eliminates potential problems when the customer picks up the car after the installation is complete.
2. Disarm the alarm system. Check for a security radio and an on-board computer.

3. Disconnect the battery, and run the power and data cables. Be sure not to slash or crimp the cables. It is important not to stress or damage either of the connectors. Be sure not to damage any existing cables.

4. Connect the power cable to ground, constant power, and switched power. Be sure to properly identify the right wires and to connect them to the correct sources.

5. Select a location and mount the antenna. Be sure all connectors are properly attached and crimped. Be sure to use the right antenna in the proper location.

6. Mount the transceiver in the appropriate location. Securely fasten the transceiver to the vehicle. Loose radio transceivers can be damaged easily.

7. Mount the control head in the passenger compartment. Pay special attention to location. Select a location that provides easy access and that doesn't get in the way.

8. Install the hands-free microphone. Pick a location that is near the driver's head and away from sources of noise. Pay special attention to any calibration procedures specified by the manufacturer of the phone. Uncalibrated hands-free devices perform poorly.

9. Connect the power and data cables to the cellular phone according to the manufacturer's specifications. Different phones use different cables and are designed to connect differently. Pay careful attention to the manufacturer's instructions. Make a quick inspection of the installation before turning the phone on.

10. Place a test call and verify the proper operation of the phone. Test the phone in hands-free and standard handset modes. Before approving the installation, be sure that you would be satisfied with the installation if this were your car.

11. Show the customer how to use his/her new phone. Don't try to teach the customer too much at one time. Keep the lesson basic. Try to make the customer feel good about the new purchase.

CHAPTER 7

Marine Installations

Marine installations offer a special challenge to installers. They can also present some rather unique profit opportunities, since not many cellular installation facilities offer this service. Marine installations are almost always done at the customer's location. It is, of course, rather difficult to bring a 50-foot yacht to your service facility.

Since sailboats and powerboats are designed in such a wide variety of configurations, it would be pointless to go into any great detail about installation. Thus, in this chapter, only general guidelines for marine installations will be given. It is important to remember that every marine installation is a custom job and needs to be treated as such. Most of the principles we discussed in the vehicle and rural installation chapters can be applied to marine installations. The most important thing to remember when doing a marine installation is to be flexible during the planning and installation phase of the job. Listen to what the customer wants, and be sure that you can fulfill the requests before committing to them.

Since a marine installation must be done at the customer's location, the installer needs to pay special attention to determining a fee for the service. The installation requires two people, along with a fairly complete set of tools. The installer must also be sure to calculate the amount of time it will take to drive to and from the installation site. It is usually a good idea to charge by the man-hour. If the customer complains about the fee, it is generally best to decline the job.

PLANNING THE INSTALLATION

Just like any other installation, the job must be evaluated before the work begins. Look around the cabin of the boat and get a "feel" for the layout. While all boats are different in their construction, there are some similarities. The mast (or masts) is located in the center of the boat. The cabin is located below the deck, and a small engine room can be located to the rear. Sailboats range in size from modest to immense. Their owners usually share an attachment to their boats that approaches love. This is a good thing to keep in mind when planning the installation (and when deciding how much to charge for the work).

Before beginning the work, discuss the installation with the customer. Find out where the control head is to be located. A good location is a navigation station, if the boat has one. A larger boat may have a captain's (or owner's) private state room where the control head can be mounted. Plan to locate the transceiver somewhere near the control head so that you don't have to run the data cable all over the boat. Most data cables are not available in lengths much over 7 meters (about 22 feet), so don't promise to hide the transceiver in some out-of-the-way location only to find that you don't have enough cable to do the job. It is not a good idea to splice the cable to increase its length—that will create nothing but problems.

One other word of caution: Avoid selling or installing a 0.6-watt hand-held portable telephone in a boat unless it is sold with a power booster that increases its power to 3 watts (or, in the worst case, 1.2 watts). The reason for this caution is the unusually long antenna cable run that is necessary in a boat installation. Marine installations are often required to operate at much greater distances from cell sites than standard vehicle installations, so this type of application needs as much power as possible.

MARINE ANTENNAS

Theory of Operation

Marine antennas are usually elevated-feed antennas that are packaged for use on boats. Their operation is no different than that of standard vehicular antennas. An antenna is an antenna, regardless of where it is used.

Applications

Keep in mind that the only person in the world who is more protective about a possession than a Corvette owner is a boat owner. The kinds of boats (or yachts) that attract the cellular telephone customer can cost more than several nice houses. I have seen cellular telephones installed in $500,000 boats, and I am sure there are many people who have them in more expensive vessels. Remember to pay attention to detail in these situations.

SELECTING THE ANTENNA LOCATION

There are two kinds of boats: sailboats and powerboats. The antenna location and installation are very different for the two.

Sailboats

On sailboats, mounting the antenna on the tallest mast is the best choice (see figure 7.1). Getting to the top is difficult, a bit frightening, and something that is not to be attempted in the wind. However, the customer will be rewarded by your efforts. This installation location will provide superior performance. Remember, water is generally flat. Electrical waves propagate very well over water. The water is as close to an infinite conducting surface as we will ever get. As a result, an antenna mounted at the top of a high mast will operate at some rather incredible distances from the cell site. I have experienced good, clear conversations when a boat is at a distance up to 40 miles from a cell site. The only real caution in selecting an antenna location is to stay away from radar equipment. A marine radar antenna looks like a "flying saucer." Do not mount the cellular antenna near the marine radar because the cellular telephone may be damaged by the high-powered signals that the radar generates.

Powerboats

Powerboats normally have a place for an antenna at the stern (rear) of the boat. Some boats have a location area near the upper bridge (see figure 7.2). Powerboats usually do not provide much choice in location. Just put the antenna where all the other ones are located. Cellular telephones operate at a high enough

146 MARINE INSTALLATIONS

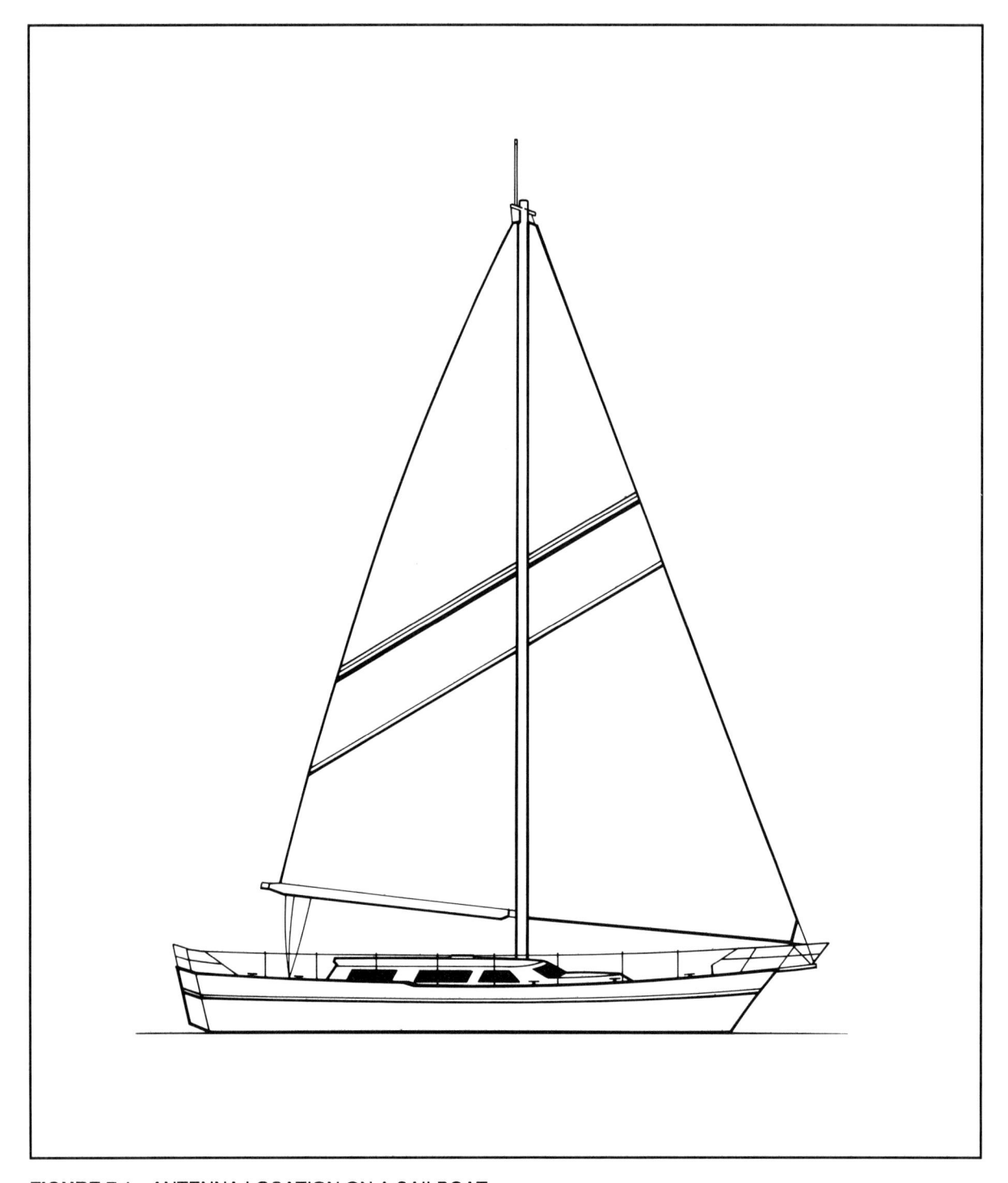

FIGURE 7.1 ANTENNA LOCATION ON A SAILBOAT

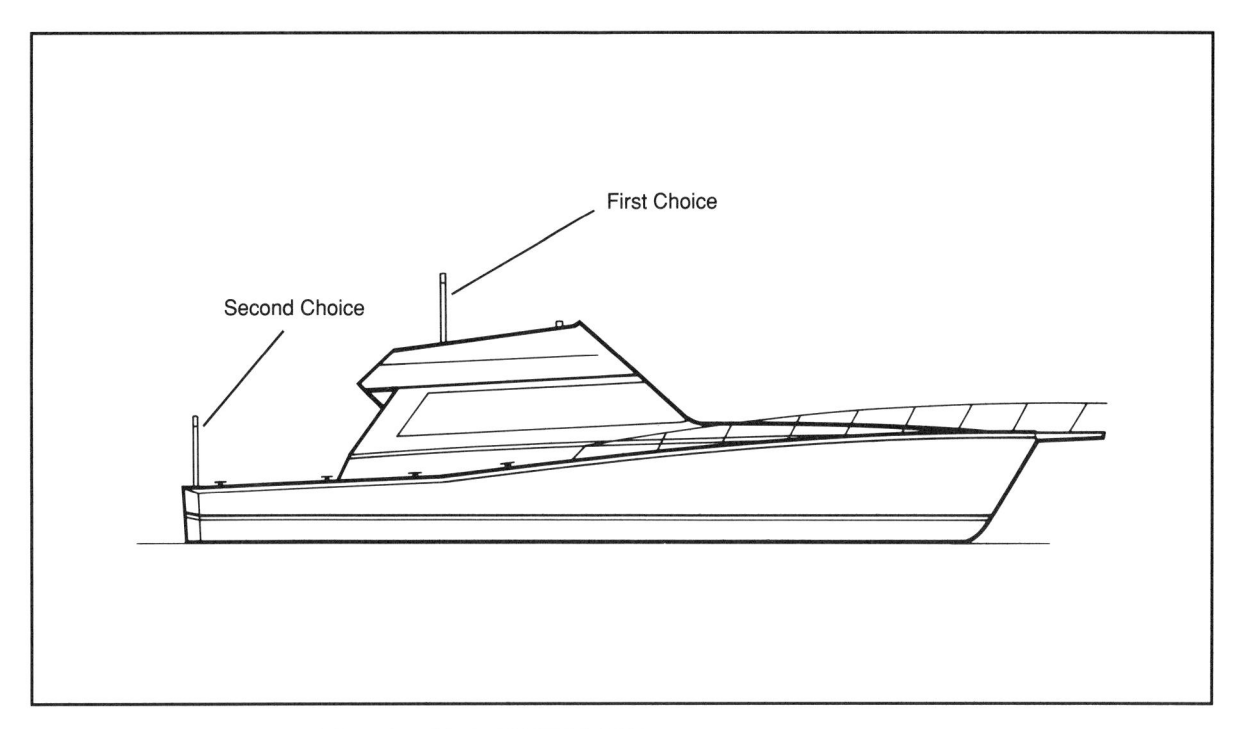

FIGURE 7.2 ANTENNA LOCATION ON A POWERBOAT

frequency so that they won't interefere with any of the navigation or communication equipment. Again, stay away from radar equipment.

INSTALLING THE ANTENNA

Sailboats

Marine installations are usually more difficult than any other type of installation. Sailboats are the worst. Remember this when pricing the installation. When doing marine work, it is a good idea to charge about twice the fee for a vehicular installation. The first thing required when mounting an antenna on a sailboat is a calm day with no waves. A boat that is pitching only a few degrees will have several feet of movement at the top of the mast. It is no fun at all if you are sitting on top of the mast while it is swinging. There is also the problem of getting to the top of the mast. Fortunately, there is usually someone at the marina who has the equipment and experience to get you to the top. The equipment consists of a small plank to sit on and a pulley arrangement to provide some mechanical advantage. There should be a safety line in case the pulley fails. Be sure to take the time to try out the safety equipment. Before ascending to the top of the mast, be sure you have all the tools you will need in your tool belt: You don't want to make the trip to the top more than once.

Once at the top of the mast (or near the top), the antenna can be mounted on a small metal shaft that protrudes from the top of the mast (see figure 7.3). The marine antenna manufacturer should provide a small clamp to do the job. After the antenna is mounted, apply some Locktite™ to the screws that tighten the clamp. This will keep the screws from becoming loose over time. Be sure to position the antenna in the vertical position and to place it as far away from the mast as possible. Fasten the coaxial cable to the metal shaft with cable ties so it won't make noise in the wind. Some boat manufacturers may have provisions for antenna cable in the mast. If not, use your battery-powered drill to make a small hole to run the cable through. Then feed the entire coaxial cable down the length of the mast. Before descending, use some silicone sealer to ensure that no water will get into the mast when it rains.

The mast runs into the keel of the boat and has a small access port at its base (in a convenient location, with some luck). Remove the access port, and pull the cable to the transceiver. Be sure to use silicone sealer to seal any holes that must be made.

Powerboats

Powerboat installations are much easier than those for sailboats. You can usually look to see where other antennas are mounted. Similarly mount the cellular antenna, and run the cable where the other cable has been run. It is usually that easy. If the powerboat does not have any existing antennas, find the highest point on the boat, and mount the antenna. Run the antenna cable the same way as was described in the sailboat section.

MOUNTING THE CONTROL HEAD

When mounting the control head, be sure to follow the customer's instructions to the letter. This may be one of the few situations where it is a good idea to have the customer with you while you are doing the installation. Since many boat owners are "tinkerers," they may know all the internal workings of the boat and be able to offer some very useful suggestions during the installation. There are a number of control head mounting brackets available

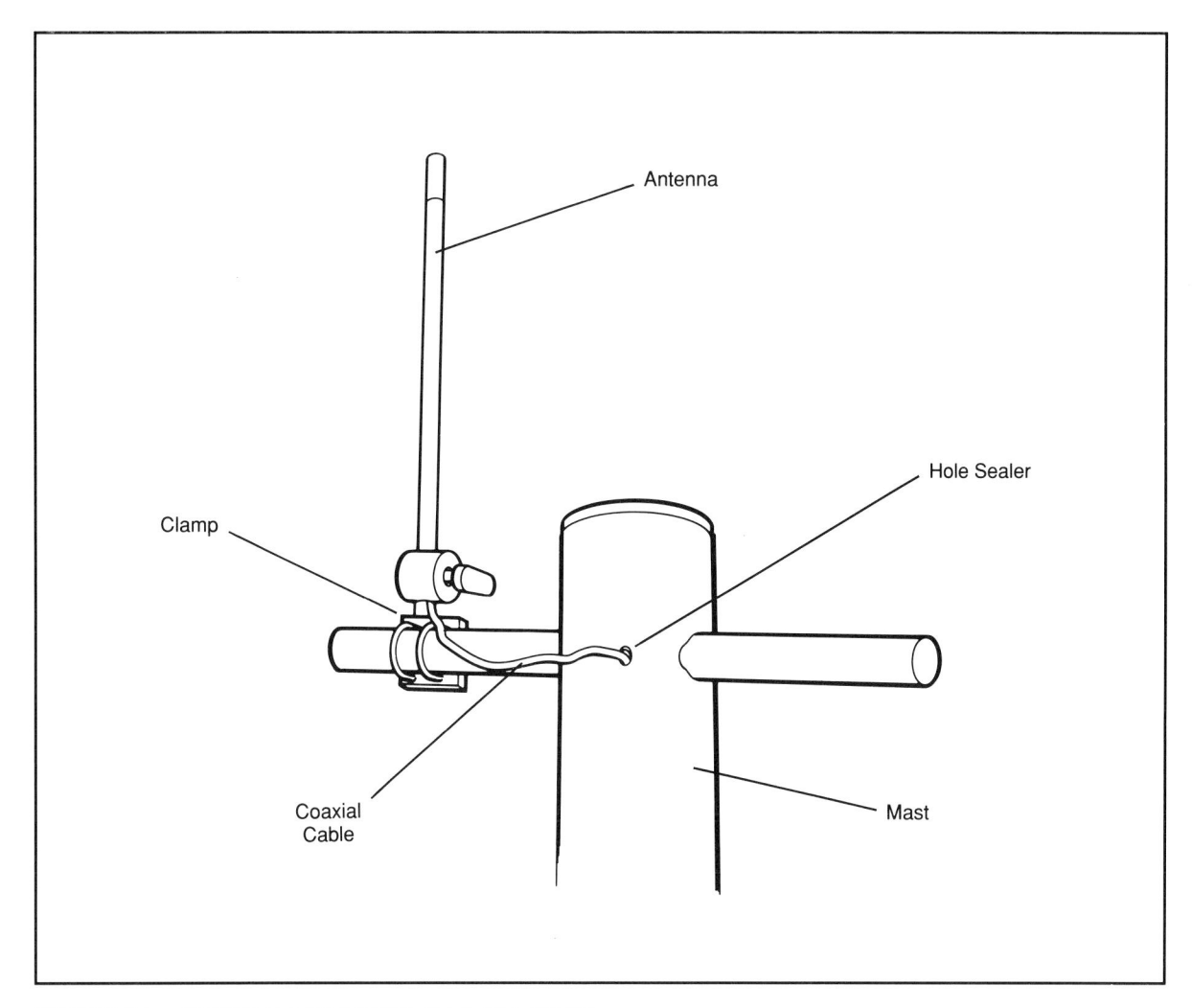

FIGURE 7.3 ANTENNA MOUNTING ON A SAILBOAT

from various manufacturers (see Appendix B, "Directory of Manufacturers"). It is a good idea to bring an assortment of them with you when doing the work. Mount the control head in an area where it is accessible but not where anyone can hurt themselves on it if the boat is in rough water.

MOUNTING THE TRANSCEIVER

The transceiver should be mounted behind a panel so that it is not visible to the customer after the installation is complete. When selecting a mounting location, be sure there is power available nearby. The last thing you want to do during a marine installation is string power all over the boat. Mount the transceiver with the supplied mounting hardware. A word of caution about the transceiver mounting: Be careful when drilling any kind of hole in a boat. Unlike cars, there are a myriad of ways to damage a boat while drilling a hole. Boats have drinking water tanks, waste holding tanks, fuel tanks, cooking fuel tanks, and assorted wiring harnesses that you will most likely not be familiar with. It is even possible to drill a hole in the boat and create a leak. Be sure that the equipment is mounted in a dry place. Don't be afraid to use a little bit of velcro during the installation process.

COMPLETING THE INSTALLATION

Look around for a good source of 12 volts. Boats have 12-volt power in many locations. Tap to the power source with standard 3M-type connectors. Some boats have a lot of external RF noise caused by diesel motors, navigation equipment, or other on-board electronic equipment. Bring a power filter to the installation location in case this problem arises. Since the customer will want to have the phone operating at all times, you will want to "hot wire" the phone to the power source. Remember to connect both the ignition sense and the constant power wires directly to a constant power source.

Connect the power cable to the transceiver and the data cable to the control head and transceiver. Attach the antenna cable to the transceiver, and turn the phone on. Using a watt meter, check to be sure that there is not much reflected power coming from the antenna. This is very important in an installation like this where the antenna cable is so long.

Take the time to show the customer where everything is mounted and how to use the phone. A marine installation will generally take longer to demonstrate than a standard vehicle installation.

ADD-ON ACCESSORIES

In some more elaborate installations, the customer may not want a "high tech" control head mounted in a traditional-looking cabin. In these cases, use an RJ11 interface in order to mount more traditional-looking telephone equipment in the cabin. Don't forget the many electronic devices that can be connected to a standard telephone line and thus to a cellular telephone. Sophisticated alarm systems can provide customers with security for their expensive investment. "Fax" machines can put them closer to their office. Video sampling systems (for the true gadget lover) can be used to send back slow-scan images of their trip as it is happening. Customers can even have several standard telephones connected at various parts of the ship for their convenience. Let your imagination work overtime and have a little fun with this type of installation. Your customer will love it. It is capabilities like these that will get you referrals for the more profitable marine work.

CHAPTER 8

Rural Installations

One of the offshoots of the cellular industry has been what is known as "fixed" cellular installations. This particular branch of the industry has focused upon applications of cellular technology in environments other than vehicles. It is an area of the industry that is still fairly unexplored and still offers significant profits to people who become involved with the technology.

After operators in the the large metropolitan areas were granted licenses to provide cellular service, there were still vast amounts of geography where cellular service was not yet available. The FCC is in the process of granting licenses to these rural areas. These areas are referred to as RSAs, or Rural Service Areas. While the population density of most of these areas would never support even a single service provider, traffic from nearby cities or highways passing through the areas can make the operation of a cellular system profitable.

The first application in the RSA that we will discuss is a fixed installation in a dwelling. A common scenario is a person with a summer or vacation home in a very remote area, so remote that no telephone service is available or the cost of running the lines to the house is prohibitive. Many of these areas don't even have electricity. All of the appliances are run by natural gas (even the refrigerator). Potential customers would like to have the ability to use a telephone when they stay at the cabin, either for business purposes or just to have the security of being able to call for help in case of an emergency. Many of the people who own these cabins are from big cities and like the isolation, but they still feel the need to be able to get in touch with the "world" at times. They just aren't willing to spend the $20 or $30 thousand that the phone company would charge to run the telephone lines to the cabin. A fixed cellular installation can be a cost-effective alternative to standard landline telephones.

Just a few tools and a little bit of imagination are required for a fixed installation in a remote location. The equipment you will be installing is very similar to that which you have been installing in cars, with a few additions. A rural installation requires the following:

1. Cellular telephone
2. Antenna with cable
3. Solar panels (photovoltaic cells) if no electricity is available
4. Voltage regulator
5. Lead-acid storage batteries
6. RJ11 interface (optional)
7. Computer interface (optional)

CELLULAR TELEPHONE

The telephone that you and your customer select can be almost anything you want. If he/she plans to install the phone system and leave it, a standard cellular telephone will be fine. Whatever you and your customer select, it will most likely have more features than his/her regular telephone at home. After all, how many home telephones have memories, alphanumeric capabilities, automatic redial, etc. Most standard home telephones only make and receive calls.

If the customer plans to use the same phone that is used in the car or at other locations, a transportable may be a good solution. It can be removed quickly from the car and installed in the cabin. All the numbers that are stored in the car phone will then be available to the customer when in the cabin. It will also save the customer the cost of activating another telephone number and paying two monthly fees for service.

I don't recommend using a hand-held portable phone in a rural location. Hand-held portables have a power output of 0.6 watts. While this power output is usually sufficient for proper operation in mature urban markets, new rural markets will have fewer cell sites and will not be covered as well. This means that it will be more difficult to place and receive calls in these new markets. Using a full 3-watt cellular unit will yield the best performance results.

ANTENNA

This equipment selection is a bit more tricky than for a cellular telephone. An installer can pick any phone, and it should work well in the cabin. The antenna is quite another story. Before selecting an antenna, you need to determine the level of the signal available in the area where the installation will be done. In many new rural systems (at least during the early phases of operation), the cellular coverage will be spotty. You should call the cellular service provider to find out how good the coverage is in the area where you plan to install the phone. After getting the information from the carrier, drive to the location and try to place a few calls from your car. Bring a phone that has a good signal-strength indicator, so you can get a quantitative idea of the signal strength that you will have to work with. If the cabin is not accessible by car, take a transportable phone and get to the cabin by whatever means necessary (helicopter, horse, canoe, etc.). Remember to charge by the hour for your time, since these installations can take a lot more time that one would expect.

If you have a good strong signal, the antenna choice is easy. A standard vehicle antenna will work fine. An elevated-feed antenna mounted on the roof of a house will usually work great. A marine antenna is also a good choice and is easier to mount since it is designed to be mounted on a flat surface (see figures 8.1 and 8.2). Be sure to select an antenna with at least 3 db of gain.

If the signal is marginal, the problem becomes more complex. There is a solution, however. Antennas with more than 3 db of gain are available for these types of installations. While their size and directionality make them unsuitable for use in a mobile installation, they work very well in a fixed environment. Several manufacturers (see Appendix B, "Directory of Manufacturers") offer yagi antennas with 9 db or more of gain (see figure 8.3). These antennas are very directional and have to be pointed. By mounting the yagi antenna on the peak of the cabin roof and pointing it in the direction of the cell site (see figure 8.4), you can obtain the extra bit of signal that will make it possible to place and receive clear telephone calls.

156 RURAL INSTALLATIONS

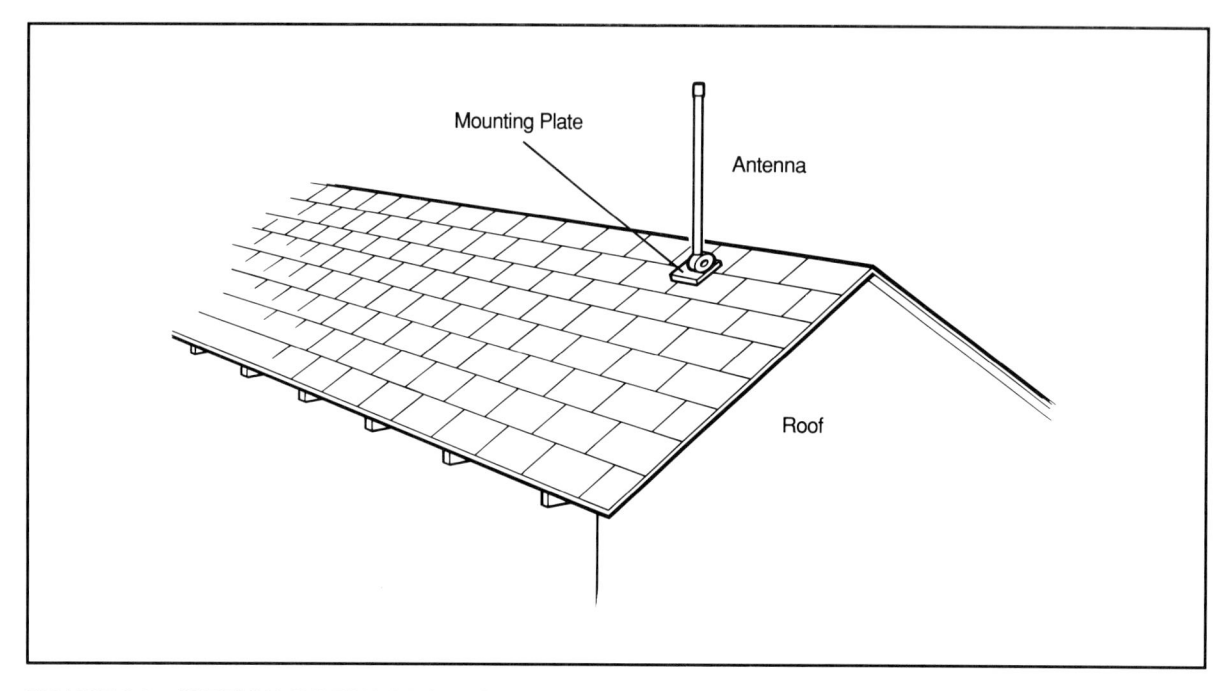

FIGURE 8.1 ANTENNA INSTALLATION FOR A RURAL CELLULAR TELEPHONE

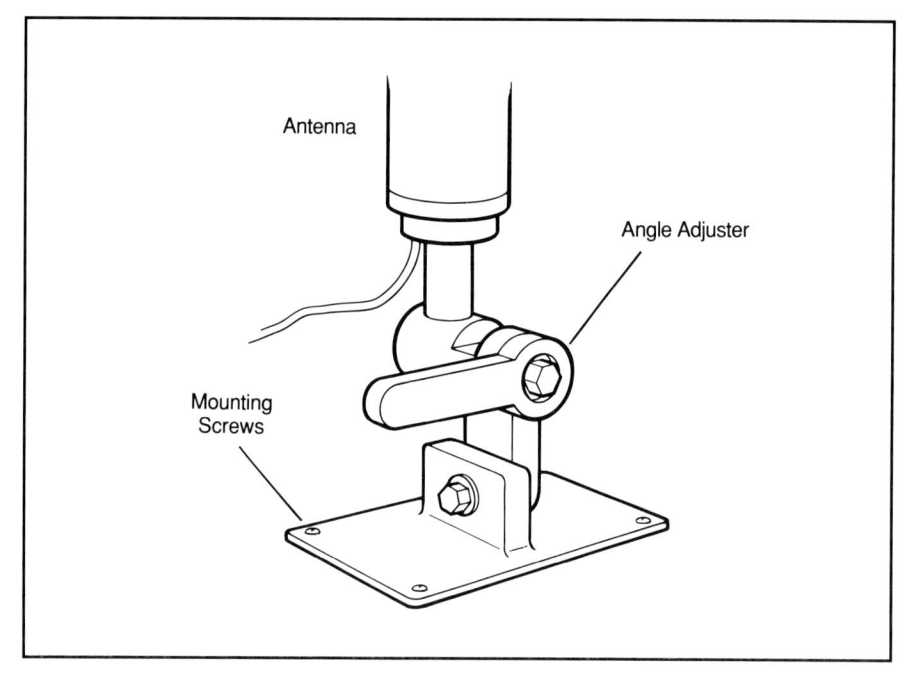

FIGURE 8.2 MARINE ANTENNA MOUNTING HARDWARE

RURAL INSTALLATIONS 157

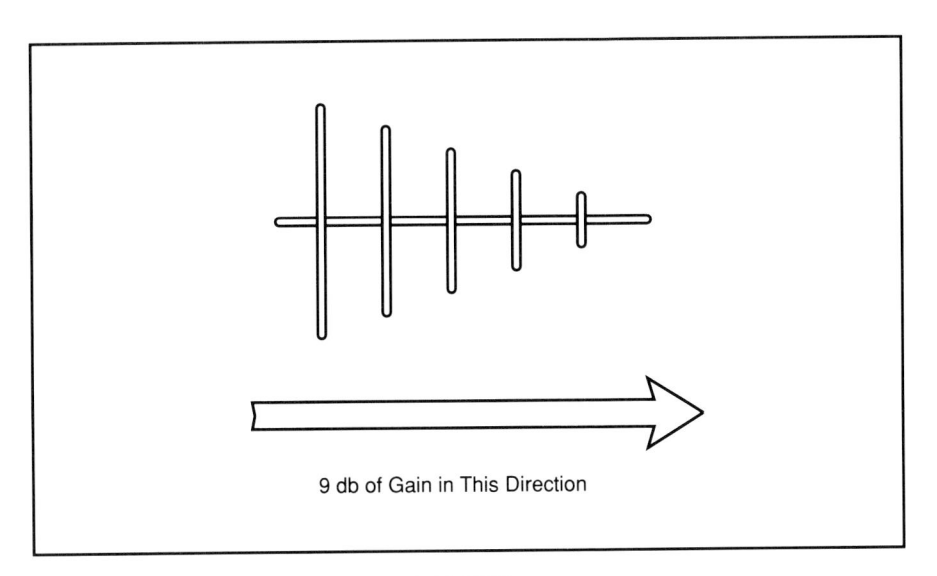

FIGURE 8.3 A YAGI ANTENNA (TOP VIEW)
A 9-bd gain antenna is designed for nonmobile applications.

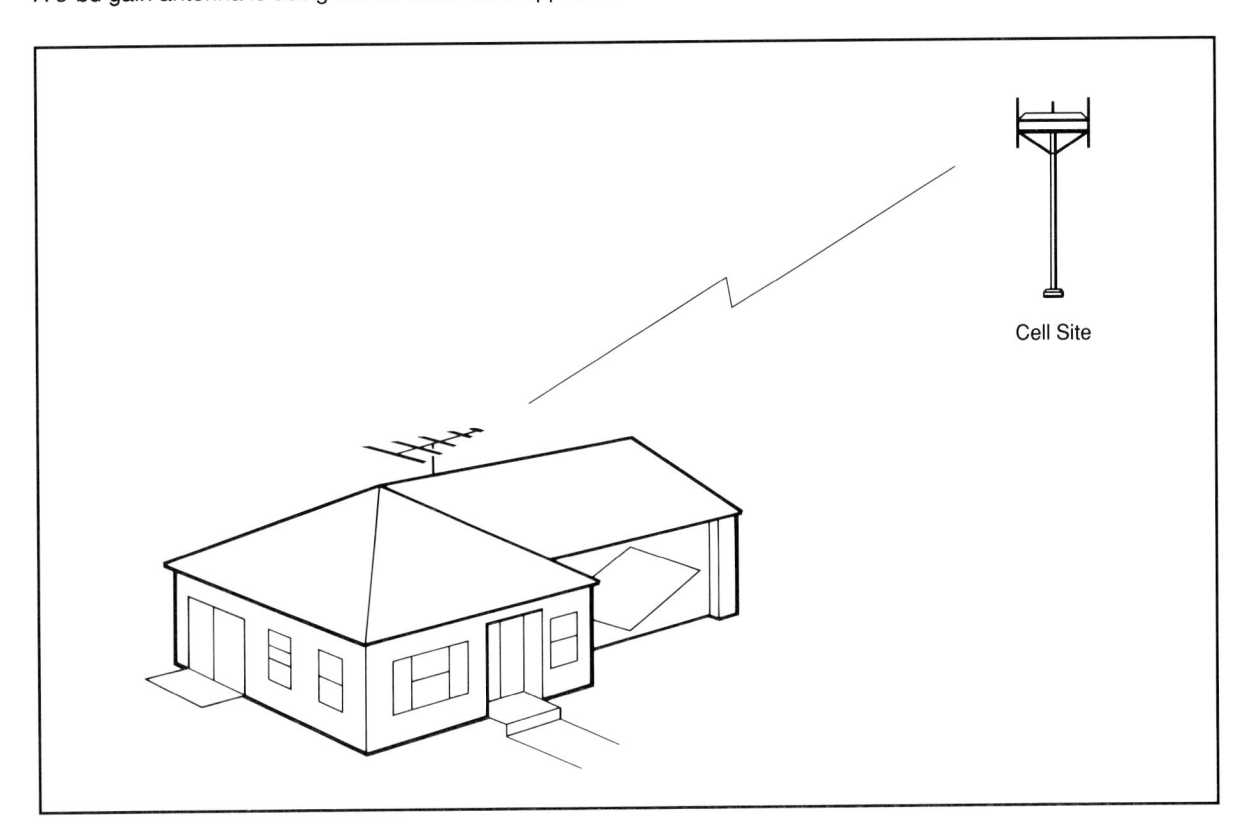

FIGURE 8.4 YAGI ANTENNA MOUNTED ON ROOFTOP

If no signal exists, you are simply out of luck. No antenna in the world will allow calls to be placed in an area where there is no signal.

SOLAR PANELS

Most remote cabins don't have electric service. If they did, it would be a simple matter to get telephone service, and the customer wouldn't need a cellular telephone. Providing power is an area where you can get creative. Many companies sell small portable electric generators that run on gasoline. They are usually priced around $1,000 and are quite reliable. Unfortunately, they are noisy and need to be started and stopped. They also have to be refueled. If the cabin is in a very remote location, obtaining the fuel can be an inconvenience.

A good solution to this problem is the use of photovoltaic cells, commonly referred to as solar batteries. These cells convert sunlight to electricity (see figure 8.5). They have no moving parts,

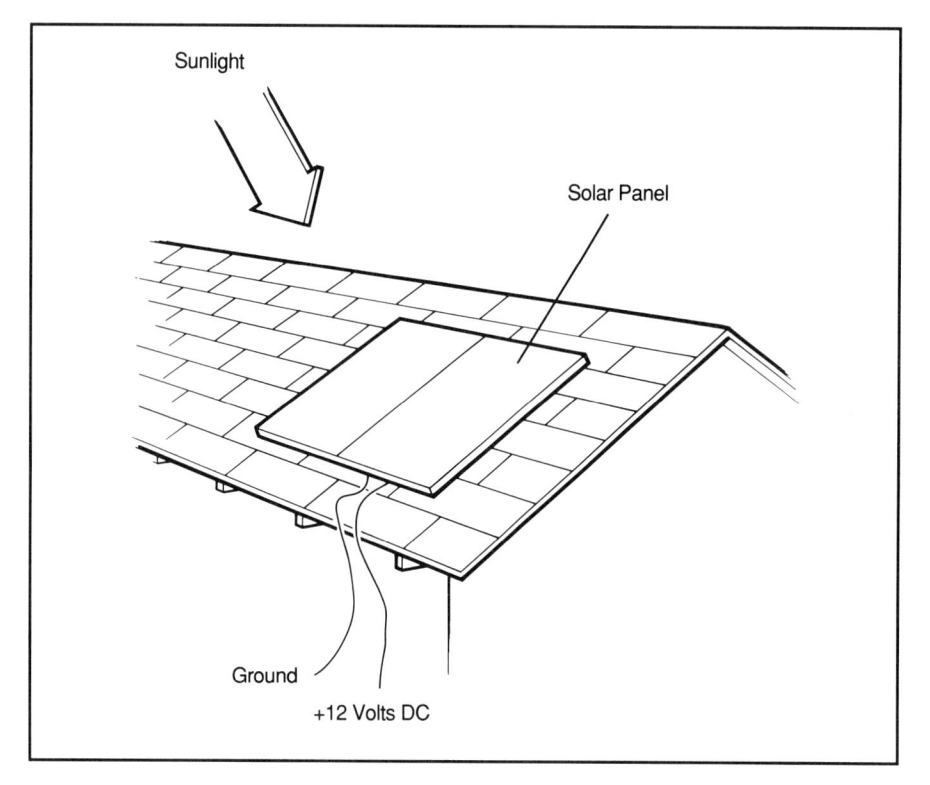

FIGURE 8.5 SOLAR PHOTOVOLTAIC PANEL
Solar photovoltaic cells can provide power for rural cellular telephones.

don't make noise, don't burn fuel or pollute, and last almost forever. These devices have been available for many years, but they have always been expensive. During the last few years, however, the cost of these cells has dropped considerably as their efficiency has improved dramatically. A few hundred dollars will purchase enough solar cells to operate a cellular telephone. When selecting a solar panel, it is a good idea to purchase one large enough to supply 12 volts at 2 to 3 amperes (at the output of the voltage regulator) in full sunlight. Refer to Appendix B, "Directory of Manufacturers," for suppliers of solar cells.

VOLTAGE REGULATOR

Unfortunately, the voltage output of a solar cell varies with the amount of sunlight to which it is exposed. This will create a problem with most cellular telephones, since they are designed to operate on a fairly constant voltage supply. The manufacturers of solar panels understand this problem, since most electrical appliances also have narrow voltage requirements. The voltage regulator, shown in figure 8.6, is designed to keep the voltage at its output terminals at a constant 12 volts, regardless of the input voltage (see figure 8.7). The manufacturer of the solar panels will offer a special voltage regulator along with the solar cells for under $100. The regulator will also protect the solar panels from being damaged by a malfunctioning battery system and will shut off the charging system when it is dark.

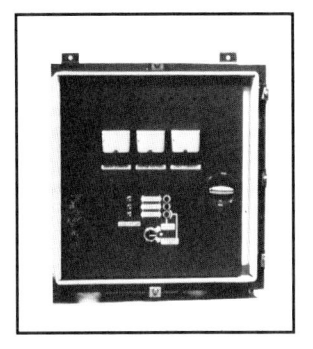

FIGURE 8.6
A VOLTAGE REGULATOR

BATTERIES

Solar panels are a nice, technically good solution to the power problem, but when the sun goes down and you want to make a phone call, what do you do? You can't store sunlight, but you can store the electricity that the solar cells generate during the day. The simplest way to do this is with a regular car battery. If the installer wants to get "fancy," a marine deep discharge battery will last a bit longer and will tolerate being completely drained a little better than a car battery. If the installer purchases two standard batteries and connects them together in parallel, they will supply enough energy to operate the phone for several days with no sunlight. This is more than enough storage capacity; even on a cloudy day, the solar panels should produce enough voltage to charge the batteries, although a bit slower than on a sunny day.

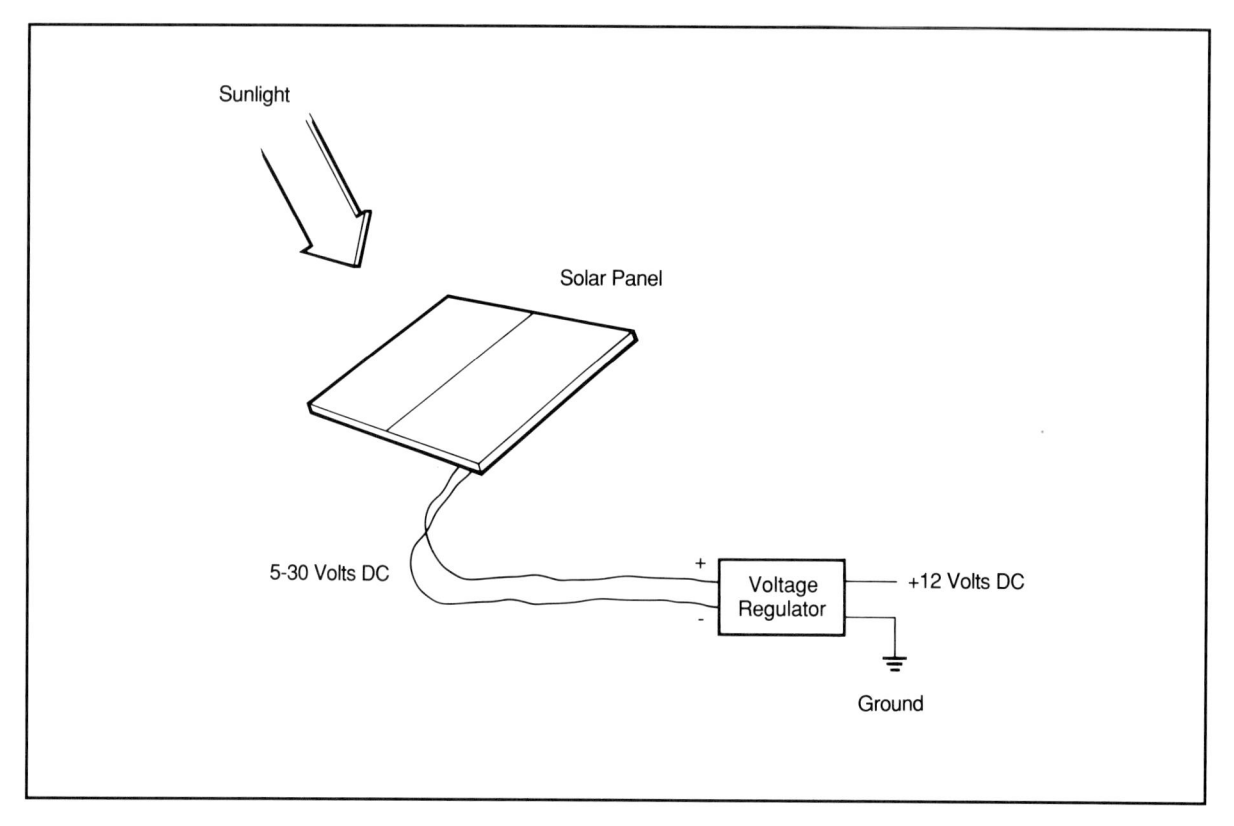

FIGURE 8.7 BLOCK DIAGRAM OF SOLAR POWER GENERATOR

RJ11 INTERFACE

Imagine you own a nice rustic cabin somewhere in the Pacific Northwest. The last thing you want is a "high tech" telephone in the interior. Fortunately, there are a few manufacturers that offer RJ11 interfaces. These devices connect to a cellular telephone (usually along the data bus) between the control head and the transceiver (see figure 8.8). Once again, see Appendix B, "Directory of Manufacturers," for suppliers of this equipment. This interface "mimics" the standard RJ11 telephone jack in your home. Anything that plugs into a standard home telephone jack will work with this interface. This means that you can take a POTS telephone and connect it to the cellular telephone. Some of the interfaces will even work with a rotary-style (dial pulse) phone. If a customer wants to retain the look of an "old" cabin, you can connect a vintage 1950s telephone to the interface and hide the "high tech" equipment out of the way.

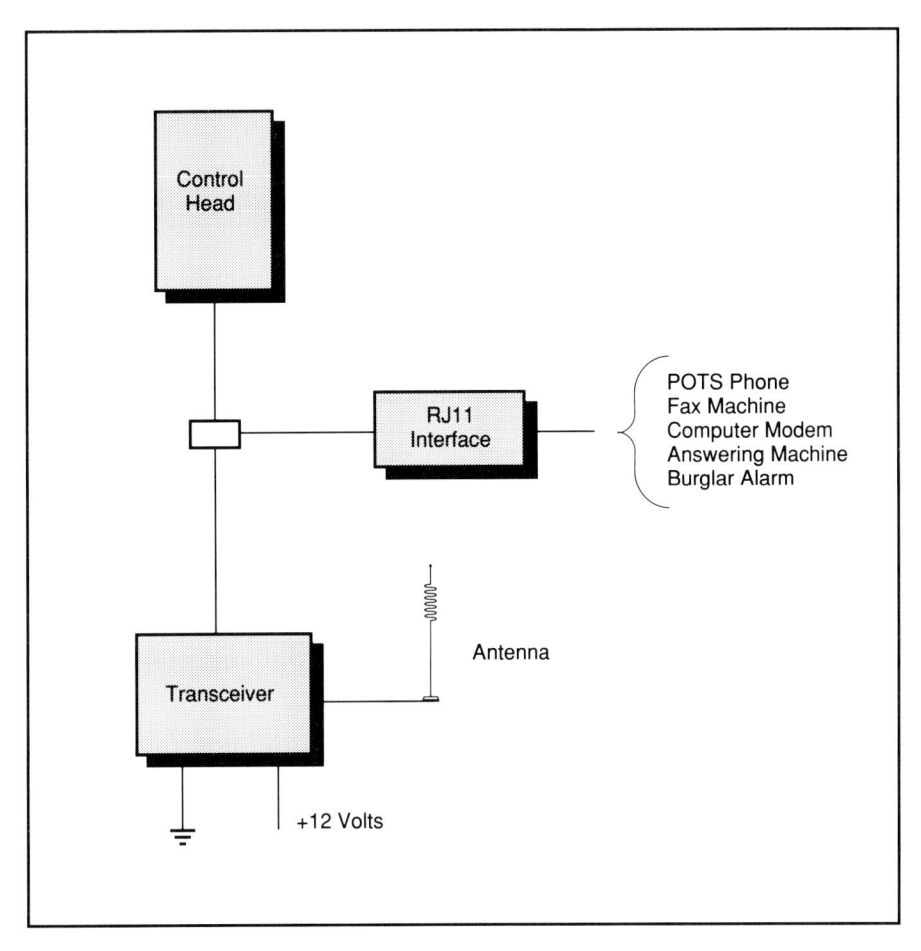

FIGURE 8.8 BLOCK DIAGRAM OF AN RJ11 INTERFACE CONNECTED TO A CELLULAR TELEPHONE

The applications of an RJ11 interface only start with telephones. A cabin in a remote area is always subject to vandalism. Installing a burglar alarm doesn't do much good if no one is around to hear it. With this interface, the alarm can be connected to the cellular telephone and programmed to dial the local sheriff in case of a break-in. The alarm system can be installed so that it is completely out of sight and does not interfere with the operation of the cellular telephone.

What about today's on-the-move executive? It is finally time for that long overdue vacation. The family is packed and ready to go to the cabin for a week. The trip has been planned for six months. The only problem is that the big takeover is behind schedule and the financial information is not going to be ready for review for two more days. Send it by express mail, right? No problem, except that "express" in a location this remote is two weeks. No need to cancel the vacation, though. Simply borrow a battery-powered "fax" machine, and connect it to the RJ11 interface. When the information is available, it can be "faxed" to you for review and then "faxed" back to the office.

COMPUTER INTERFACES

For the customer who needs to be in touch all the time in a "high tech" way, a unique computer interface is available. Spectrum Cellular has one of the few products that I will recommend in this book, since they offer a "one of a kind." Spectrum manufactures error-correcting cellular modems called the *Bridge*™ and the *Span*™ (see figure 8.9). The Bridge connects between the control head and the transceiver of a cellular telephone, as shown in figure 8.10. The computer connects to the Bridge™ via an RS-232 port (a common computer interface). The computer can control all the functions of the cellular telephone and can actually place and receive phone calls. The Bridge™ and Span™ pair provide for full error correction during handoffs or periods of low signal. This keeps vital data from being lost or sent incorrectly. Imagine doing a bank transfer and sending $1,000 instead of $100 because of a data error.

INSTALLING THE FIXED INSTALLATION

Now that we have examined the basic building blocks of a fixed installation, let's begin the work.

Antenna

The first piece of equipment to install is the antenna. The best place to mount the antenna is at the highest point of the cabin or house. This location will provide the best signal transmission and reception. If you choose to use a marine antenna, the mounting will be quite easy. Simply pick the spot and attach the mounting plate to the roof, using either screws or roofing nails. Adjust the antenna so that it is vertical.

RURAL INSTALLATIONS 163

FIGURE 8.9 ERROR-CORRECTING CELLULAR MODEMS

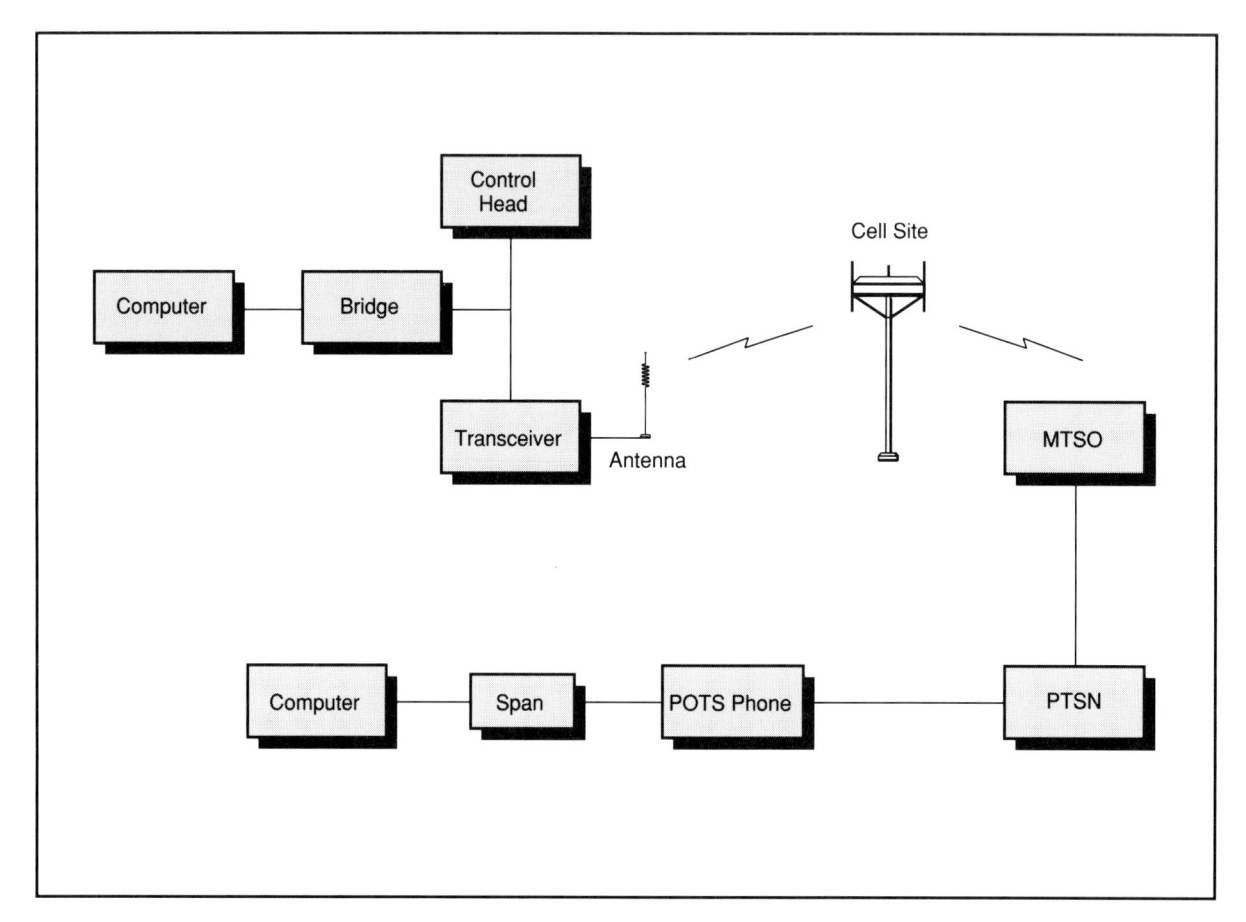

FIGURE 8.10 BLOCK DIAGRAM OF AN ERROR-CORRECTING CELLULAR MODEM OPERATION

Running the Cable

There are two approaches to running the cable. The first is to run it on the outside of the roof. Keep some roof tar handy to secure the coaxial cable to the roof. Be sure to secure it properly; otherwise, it will make quite a racket in a high wind. Run the cable along the edge of the roof, and find a convenient place to work the cable inside the building (see figure 8.11). If there are no pre-existing holes, drill a hole large enough to accommodate both the coaxial cable and the power cable from the solar panels. To prevent water from getting into the house, be sure to seal the hole with silicone window sealer after the cables have been run.

The second approach is to drill a hole in the roof right near the antenna and to run the cable into the attic. I don't favor this approach, since it is a great way to create a roof leak. If you insist

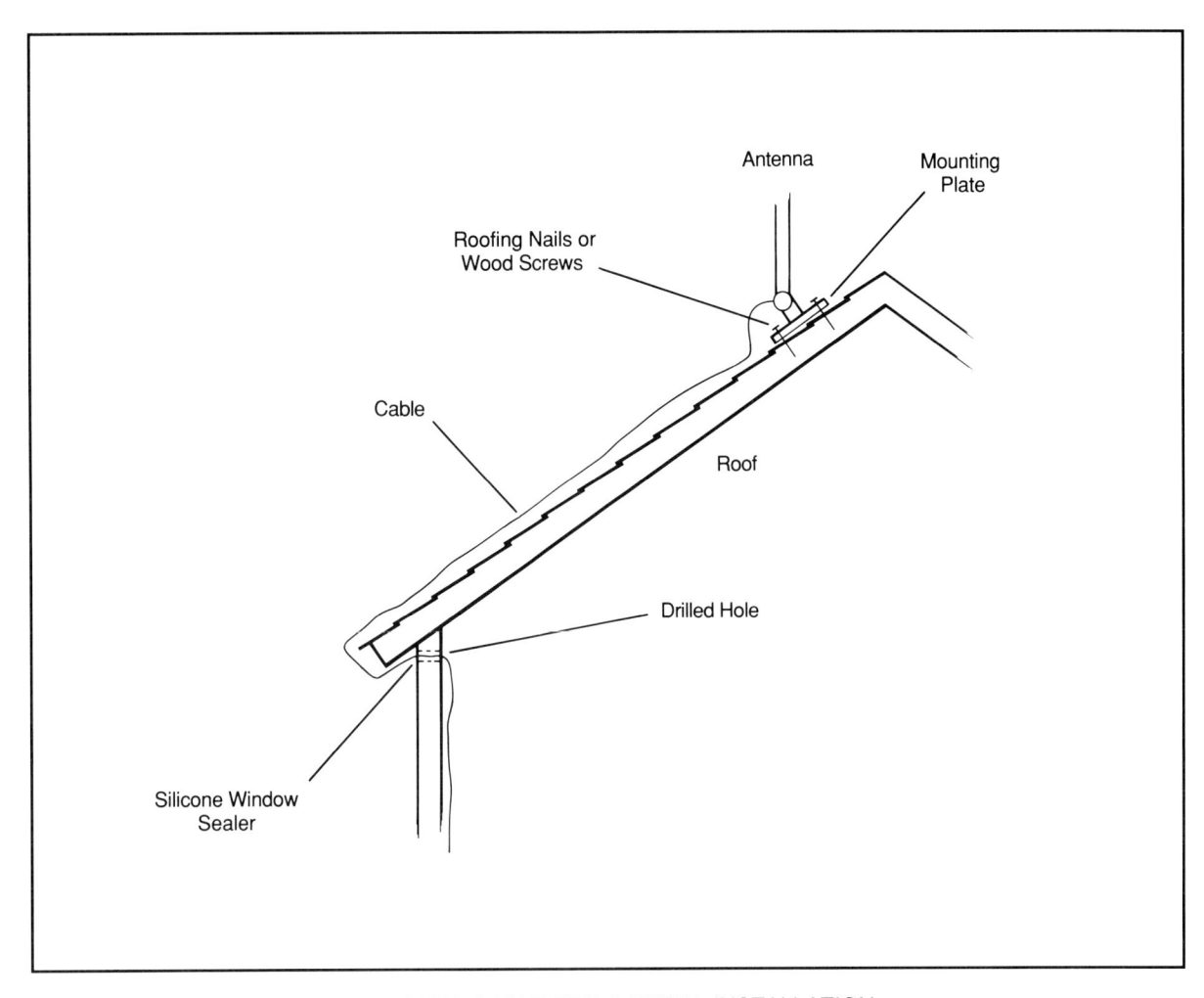

FIGURE 8.11 RUNNING THE ANTENNA CABLE FOR A RURAL INSTALLATION

on taking this approach, be sure to use plenty of roofing tar to seal the hole after the antenna cable has been run.

In either case, the cable should terminate in some kind of utility closet so that it is completely out of the way. A closet is also a good place to house the transceiver and batteries. As described in Chapter 6 (pp. 126–27), attach a coaxial connector to the end of the cable, and leave the cable for the time being.

Solar Panels

The solar panels are very easy to install. They will be in a prepackaged housing that is ready to mount on the roof. As shown

166 RURAL INSTALLATIONS

in figure 8.12, the solar panel is mounted on the roof with roofing nails. The panel is shipped with mounting brackets, so you only have to nail it to the roof. Be sure to cover the roofing nails with roof tar to ensure that there will be no leaks. Run the power cable next to the antenna cable and through the same hole so that it terminates in the utility closet.

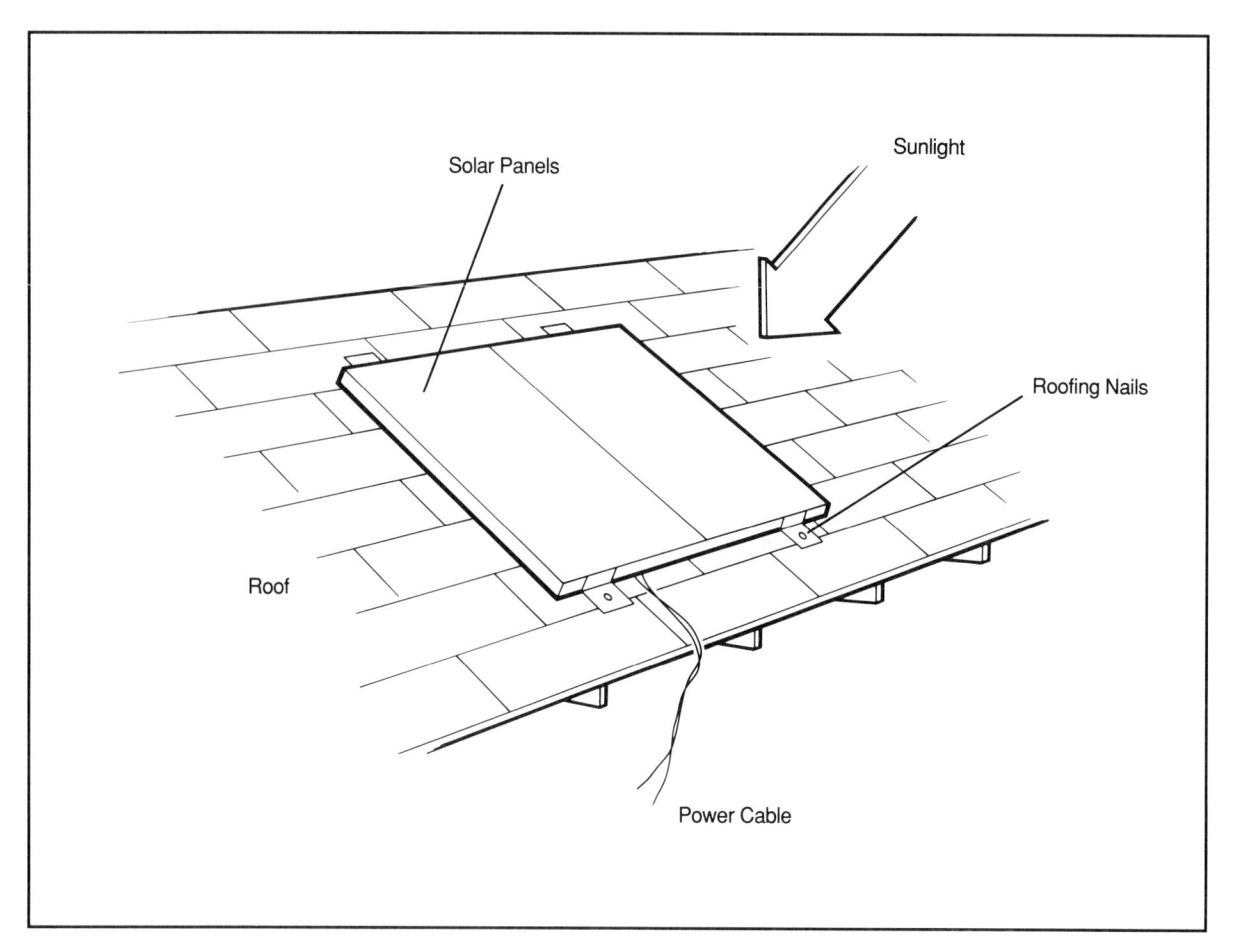

FIGURE 8.12 SOLAR PANEL MOUNTED ON ROOF

Utility Closet Equipment

Once the solar panels and antenna are mounted, you can begin to install the equipment in the utility closet. As shown in figure 8.13, the voltage regulator and transmitter are mounted on the wall. The best way to accomplish this is to first mount them on a piece of plywood and then attach the plywood to the studs in the closet.

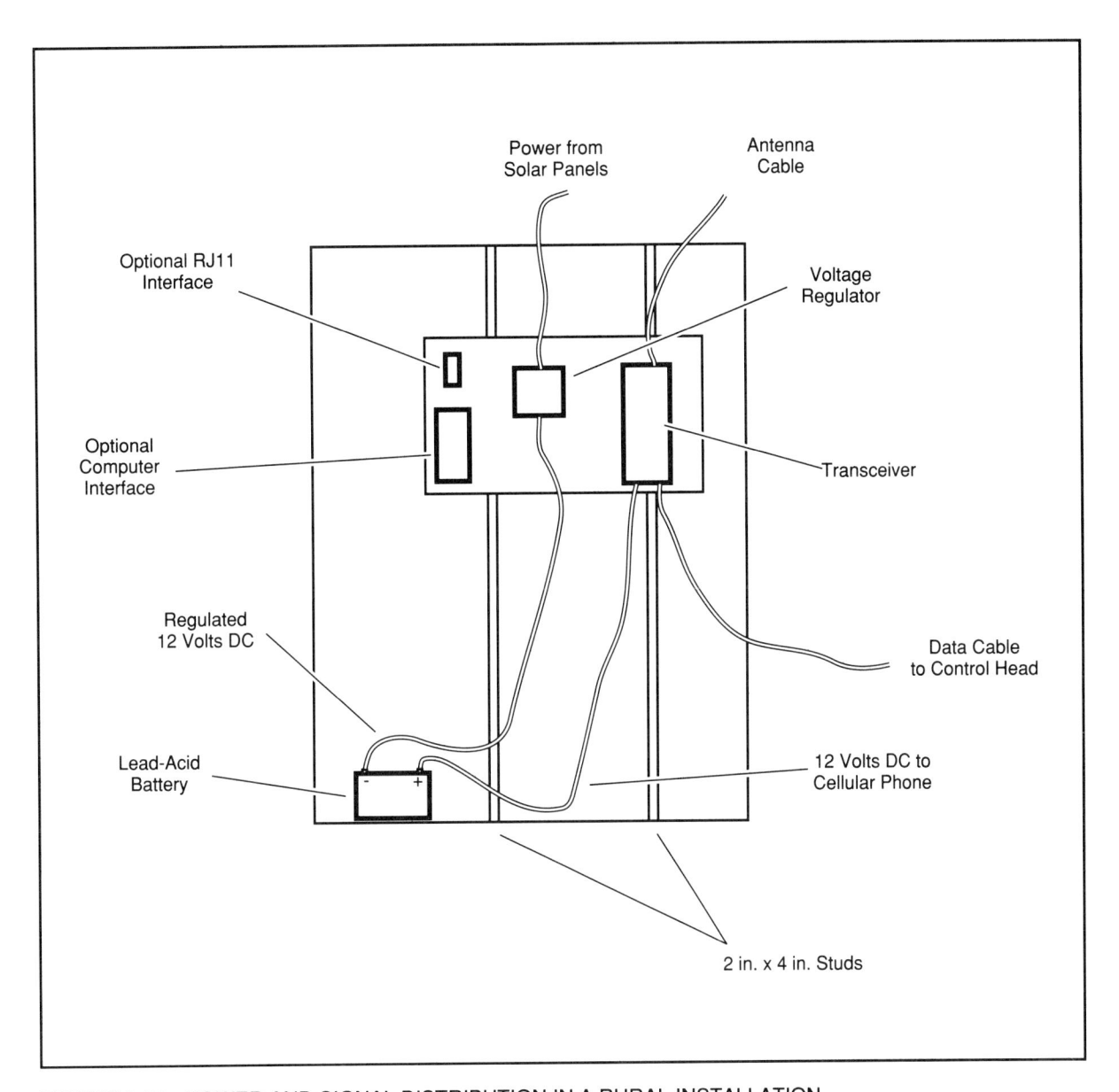

FIGURE 8.13 POWER AND SIGNAL DISTRIBUTION IN A RURAL INSTALLATION

The power cable from the solar panels is connected to the voltage regulator. The regulator ensures that the power reaching the storage battery is constant. The output of the regulator is connected to the storage battery, where it charges the battery. Even when the panels are exposed to direct sunlight, they are not really running the phone: instead, they are charging the batteries.

While this may not be the most efficient way to run the phone, it ensures that there will always be power to operate the phone. The cellular phone is then supplied with power by connecting the ground cable to the negative (−) terminal on the battery. The "hot" lead and ignition sense cable are both connected to the positive (+) terminal of the battery. Refer to Chapter 6, pp. 122–23, for instructions on how to connect power cables to a vehicle battery.

The data cable is then run to where the control head is going to be mounted. Be sure to keep the cable out of sight. A word of caution about running cables: Older cabins tend to have mice, and mice love to chew on things like data and power cables. If the installation stops working after a year or two, look for cables that have been chewed on. While there isn't much you can do about the mice, it's a good idea to be aware of their existence. The control head can be mounted using standard vehicle mounting hardware described in Chapter 6, "Vehicle Installations."

If the customer wants something special, such as an RJ11 interface, it can be mounted in the equipment closet. The "phone" lines are then run into the room where the phone is going to be located. Several extensions can be run from a single interface. A person using the POTS phone does not even need to know that a cellular telephone is being used.

The computer interface can also be located in the equipment closet on the same plywood panel as the phone and voltage regulator. The RS-232 cable can be run to a more convenient location. RS-232 cables should not be run more than 25 or 30 feet; otherwise, problems will occur.

One word of caution about lead-acid batteries: They release small amounts of hydrogen gas while charging. Because hydrogen is explosive, don't smoke around the batteries. Under certain unusual circumstances, enough hydrogen could build up in a small space to ignite if someone were smoking.

CHAPTER 9

Troubleshooting and Repairing Installations

One day a customer drives into your shop and announces that the phone you installed last year no longer works (or better yet, the phone your competitor installed). He wants it fixed while he waits. Cellular is one of the few industries where customers do not expect to leave a product for repair. They have become dependant upon their cellular phones and don't want to be without them for even one day.

Another scenario: You have just completed one of your best installations in a brand new automobile. The installation looks so good that the customer is going to think the work was done at the factory. Everyone in the shop is standing around and admiring your work. Unfortunately, the phone doesn't work.

Don't feel badly. This sort of thing happens to every installer. This is the part of the job that separates the professional from the amateur. The most important concept an installer can learn about troubleshooting problems is how to properly identify them.

Identifying problems is a logical process that requires information gathering, thought, and action. The first part of the process is gathering information from the customer. When a customer says that his phone doesn't work, what does he mean? Find out *how* it doesn't work. Later in this chapter, we will discuss some common problems with cellular telephones and some of their solutions. These solutions don't work all the time, and some problems can be rather tricky to identify. However, with a little bit of practice, you can become an expert in cellular telephone repair. And, while repairing cellular telephones may not appear to be very profitable on the surface, it is a great way to win new customers. How many times have you heard someone brag about the great person he found to repair his car or television? With a little bit of good work, it could be you that he is bragging about.

Before discussing the more common problems that occur with cellular telephones, we need to cover one more topic. This portion of the book is not intended to replace service training from the manufacturers of cellular telephones. Each manufacturer's products have subtle differences. Some manufacturers allow troubleshooting to the circuit board level, while others require that the entire defective transceiver or control head be returned for repair. It is not the intent or within the scope of this book to make the installer an expert on service procedures for every piece of equipment on the market today. The best way to learn the troubleshooting techniques of each manufacturer is to attend their service training school.

CELLULAR TELEPHONE SUBSYSTEMS

Cellular telephones consist of a number of subsystems or modules. Before troubleshooting a defective cellular telephone, let's discuss its subsystems. A cellular transceiver is usually comprised of the following:

1. Power board
2. Logic board
3. Transmitter board
4. Receiver board
5. Synthesizer board
6. Control head board
7. Audio board

Some cellular phones are configured differently, and some have several of these functions combined into one board. Let's review the functions of each of these boards.

1. POWER BOARD. This board provides electrical power to all the other parts of the cellular telephone. It is responsible for providing power at the correct voltage levels and for filtering out any unwanted signals (noise).

2. LOGIC BOARD. This board contains the microprocessor that performs all the "housekeeping" for the cellular telephone.

The logic board is usually responsible for things like hand-offs, internal communications for the phone, levels, and channel assignment. Logic failures are probably the most common type of failure for a cellular phone.

3. TRANSMITTER BOARD. This board is responsible for converting the audio information into an RF signal that is transmitted. It does this by modulating the signal. The transmitter must be capable of changing channels and power levels very quickly under the control of the logic board.

4. RECEIVER BOARD. This board is responsible for detecting the signals that are broadcast by the cell site and converting them into audio information. This is accomplished by demodulating the RF signals.

5. SYNTHESIZER BOARD. This board generates all the tones at the very specific frequencies required by the phone. These tones are used by the transmitter, receiver, and logic boards. It is important that the tones generated by the synthesizer board be very stable and exactly on frequency.

6. CONTROL HEAD BOARD. This board is responsible for communicating with the transceiver and controlling the functions of the control head.

7. AUDIO BOARD. This board controls the various audio signals. It is responsible for the processing that is done to the audio before it is transmitted and after it is received.

THE PHONE TURNS ON, BUT WILL NOT PLACE OR RECEIVE A CALL

Operator problems are one of the more common challenges facing a service department today. Whenever customers have problems with their cellular telephones, the first person they call is the service manager of the shop where they had their phones installed. This happens even when the problem is with the cellular system and not with the phones.

The most common service problem is when the phone turns on and appears to be functional, but it will not place or receive a call. The symptom is usually described by the customer like this:

My phone turns on with the car like it is supposed to, but I can't make or receive calls. I dial the number and press the SEND button, and all I get is a fast beep after about 5 seconds.

172 TROUBLESHOOTING AND REPAIRING INSTALLATIONS

This description usually identifies the problem as one with the carrier. The first step the installer should take is to verify that the problem is *NOT* with the installation. The worst possible thing a service manager can do is to immediately blame the carrier for the customer's problems. The service manager must be sure the customer's equipment is working before assuming the problem is with the carrier's system.

The customer's equipment can be checked in two ways. The first method is to remove the phone from the car and connect it to a cellular test center (see figure 9.1). Have a programmer run the auto test to ensure that all of the phone's components are working properly. Make sure the phone will both place and receive calls properly.

The quickest and easiest way to test the operation of the phone is to try to place a call on the alternate carrier's system.

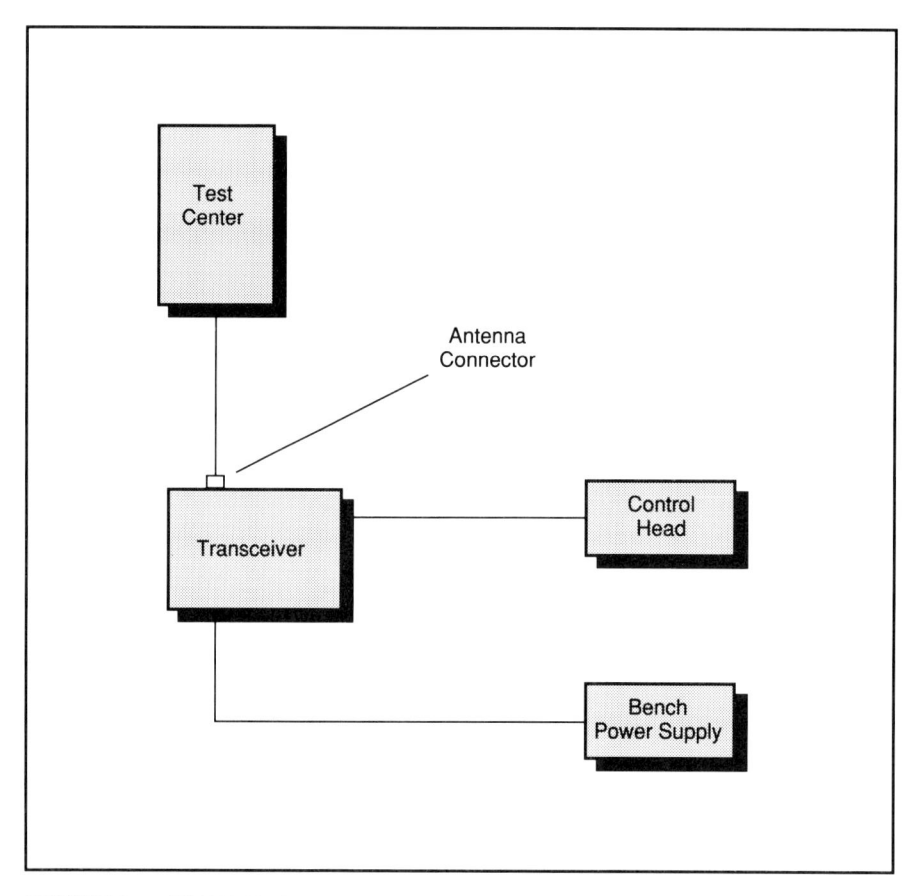

FIGURE 9.1 TYPICAL CELLULAR TEST CENTER CONNECTIONS

Simply use the A/B switch to change from the preferred carrier. Remember, in most major cities, two different carriers provide cellular service. The customer has the choice of subscribing to either of the two carriers. If the phone is working properly, the alternate carrier will usually return an announcement that is intended to let the customer know that he/she is not registered as a subscriber on their system. Although you won't be able to have a conversation with anyone to confirm operation of the audio portion of the phone, it is a very quick and easy way to verify proper operation of the call processing functions of the phone. While the call is being routed to this announcement, all of the phone's call processing functions must be used. If the call is routed to this announcement, it is probably working properly. If a call will not process on the alternate carrier's equipment or on the cellular test center, the problem resides with the equipment, not with the carrier. The phone will then require repair, and we will discuss this procedure later in the chapter.

If the phone *does* work on the cellular test center or on the alternate carrier's system, the problem may be with the carrier with which the customer is registered. In this case, the installer needs to ask the customer more questions. The first questions should be the following:

Was the phone ever able to place calls? Did the phone just recently start having this problem? Are there some areas where the phone will place calls and others where it won't?

These are very important questions with equally important answers. If the phone has never been able to place calls (this happens most often with recent installations), the problem is most likely registration of the phone number with the cellular service provider. Each cellular telephone is assigned a unique 11-digit (decimal or base 10) electronic serial number (ESN). When the phone is programmed with a cellular phone number, the phone number is "matched" with that particular ESN. The cellular carrier has both the cellular phone number and ESN stored in a data base. Every time anyone attempts to place or receive a call, the MTSO checks with the data base to be sure that the customer is a valid one. This furnishes the subscriber with quite a bit of security, since even if someone programs another phone with his/her cellular telephone number, the ESN will not be the same and the cellular carrier will not provide the "illegal" phone user with service. The installer should never give a customer's ESN to anyone but the customer.

When a customer cannot make a call (and has never been able to make one), it is often because the ESN has been programmed into the MTSO incorrectly. Since a number of people handle these numbers, it is easy for someone to switch two digits, copy a number incorrectly, or miskey a digit in the ESN. The installer may incorrectly read the serial number from the transceiver, or the manufacturer may even mislabel the phone. When a customer is denied access to the switch, these are the kinds of mistakes to look for. It doesn't help the situation to verbally abuse the cellular carrier if they are responsible for the mistake. Sooner or later, you or one of your installers will make the same mistake.

If the ESN checks out properly with the cellular carrier, verify once again that the right number was programmed into the phone. This should be done on a cellular test center—and checked carefully. Don't trust anyone to do this for you: Verify the proper programming on a cellular test center. (A test center will display both the phone number and ESN.)

If all of these items check out, there is one more possibility to consider. All cellular carriers maintain a list of ESNs in a *negative directory*. These are usually serial numbers of stolen units, delinquent customers, or other groups of customers to which the carrier does not wish to provide service. Cellular carriers are also practicing *dynamic subscriber verification* with roamers in adjacent areas. Here is how it works:

A roamer from Stamford, Connecticut (a Lynx customer) drives to New York City. When in Manhattan (a NYNEX service area), the customer places a call. While the call is being processed, the NYNEX switch "calls" the Lynx switch and asks it to verify the ESN/phone number combination. If the combination is valid, the call continues. If the combination is not valid, the call is discontinued and the subscriber's ESN is recorded in the negative file.

A customer whose number is in the negative directory will not be granted service by that MTSO. It is worth asking the carrier to check this as a last resort.

Let's look at a variation of this problem. The customer has had the phone for some time, and it has been working. Only recently has service been denied by the carrier. This sometimes happens by mistake. The most frequent cause is late payment by the customer. Find a tactful way of asking the customer if the payment for service was overlooked or made late. Call the carrier customer service department to verify this. Many carriers are becoming very sensitive about late payments (and with good cause). Get to know the people at the carrier's customer service department. They will usually be glad to help with any billing problems because they are anxious to collect their bills and get the customer back on the air and generating revenue.

THE PHONE USUALLY PLACES CALLS, BUT NOT ALWAYS

What if customers complain that sometimes they can make calls and sometimes they can't? The installer must dig further and collect more information from the customer. How often are they denied service and where? If it happens only at a certain time and location, it could be poor coverage or a cell site that is operating at capacity. Become familiar with the RF coverage of your area. Cellular phones will often not place calls in marginal areas even when their "no service" lights are off (indicating the presence of a paging channel and the ability to place a call). Cell sites (in remote areas) often operate at power levels of 100 watts ERP. A cellular phone is capable of delivering 3 watts to the antenna (6 watts ERP). There will be times when the cell site can be received because of its high power output, but the cellular phone has insufficient power to reach the cell site. In these situations, the customer will press the SEND key and will receive a "no access" tone in about 5 seconds. Take the time to learn the "bad" spots in your carrier's system.

If the cellular phone will not make a call on the alternate carrier's system, it is a safe guess that the problem resides in the phone. The cellular test center should help to identify the problem. If the manufacturer doesn't allow the installation facility to troubleshoot the problem to the board level, the phone needs to be sent back to the factory for repairs. If the service facility is allowed (and has the capability) to troubleshoot to this level, it is time to open up the phone for repair.

Read the power output from the cellular test center. If the power level is very low, attempt to adjust the level to 3 watts according to the manufacturer's instructions. If you can't adjust the power or don't have the equipment to do so, replace the transmitter board in the phone. When doing this, be sure that you are wearing a grounding strap to prevent accidental static discharge from destroying the delicate CMOS circuits in the board. After replacing the suspected board, reconnect the phone to the test center and measure the power output. If the power is still not a full 3 watts, replace the logic board. The logic board controls the power output of the transmitter and is a common cause for insufficient power output.

THE PHONE WILL NOT TURN ON

Another common problem is a phone that will not power up, or turn on. It is relatively rare for a cellular telephone to completely fail. When a phone will not turn on, the problem is usually fairly simple. The first potential problem you should check is the phone's power switch. This may sound trivial, but when a phone is wired to turn on and off with the ignition, the customer rarely touches the power switch. A passenger or loose object can accidentally turn off the power to the phone. Believe it or not, it happens. Be sure the phone is turned on.

If this is not the problem, the next thing to check is the fuses. Don't just inspect them visually—test them with a multimeter. To do this, remove the suspected fuse from the fuse holder, and attach the test probes to each end of the fuse (see figure 9.2). When the multimeter is set to ohms, it should read 0 ohms or a short circuit.

If the fuse is okay, you should next check the power line. Be sure it is securely attached to the vehicle battery. Ground the proper end of the test light, and touch point 1 as shown in figure 9.3. The light should illuminate, indicating the presence of 12 volts. If it does not light, the problem lies between the car battery and the fuse. However, if the light indicates the presence of 12 volts, check the other side of the fuse (point 2). The light should once again illuminate. If it does not light, you know that either the fuse is blown or the fuse holder is defective.

If this part of the circuit tests okay, check the connector that plugs into the transceiver. If there is no power here, you know the problem lies between the connector and the fuse connector.

After checking this part of the power circuit, be sure that the ground wire is providing a good, solid ground. Inspect the ground point to be certain that there is a good mechanical connection. Corrosion or a loose connector can keep the transceiver from being properly grounded (see figure 9.4). This connection can corrode or loosen over time. A good way to prevent this from

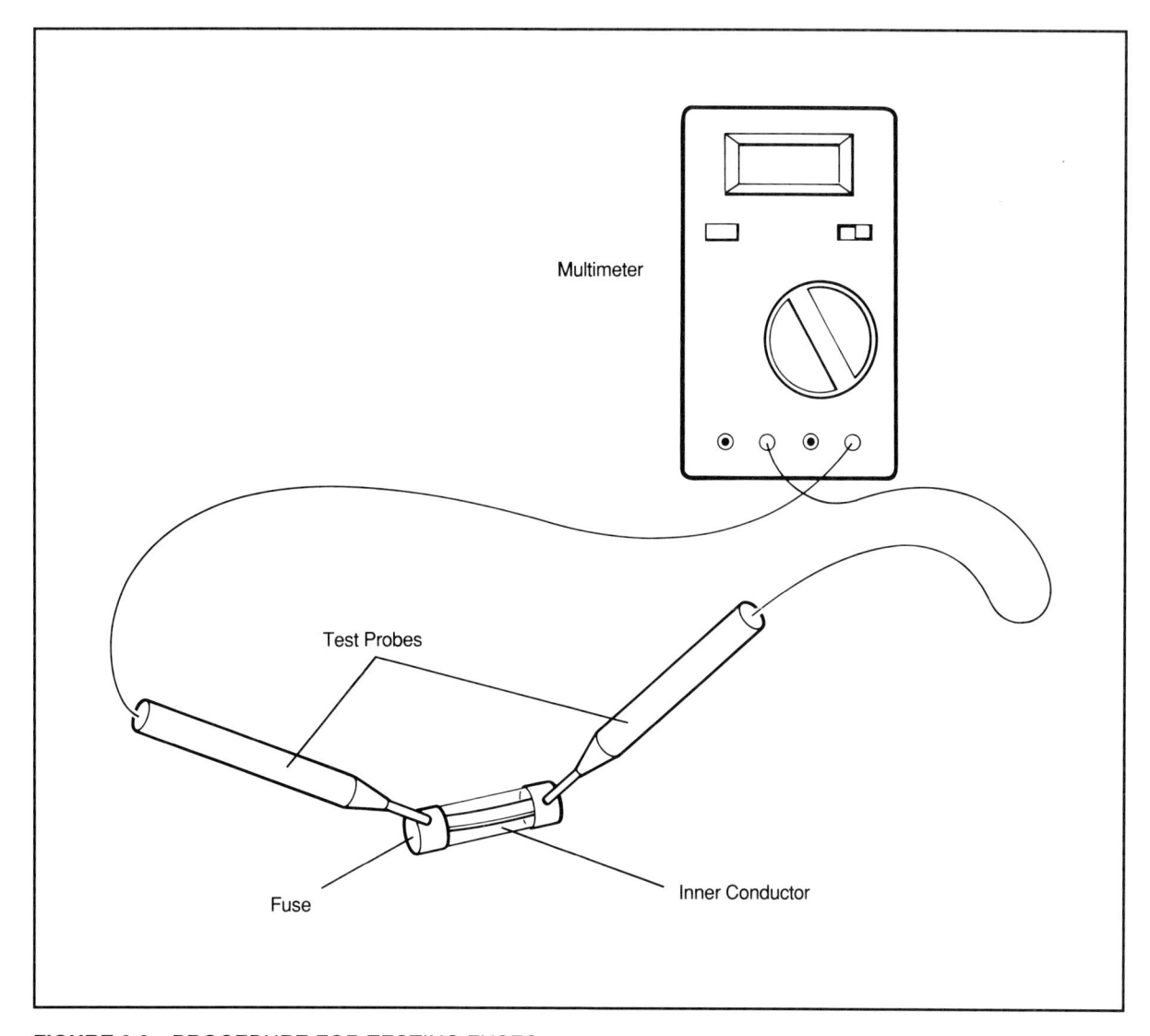

FIGURE 9.2 PROCEDURE FOR TESTING FUSES

happening is to use silicon sealer to protect the bare metal when doing the initial installation or a repair. Without a good electrical ground, the cellular phone will not power up.

Next, check the ignition sense line. Place the probe of the test light or multimeter inside the connector on the ignition sense line. This will vary from manufacturer to manufacturer, so be sure to check the service manual for the location. While doing this, have another installer turn the ignition on and off. The meter or test light should show the power being turned on and off. If no power

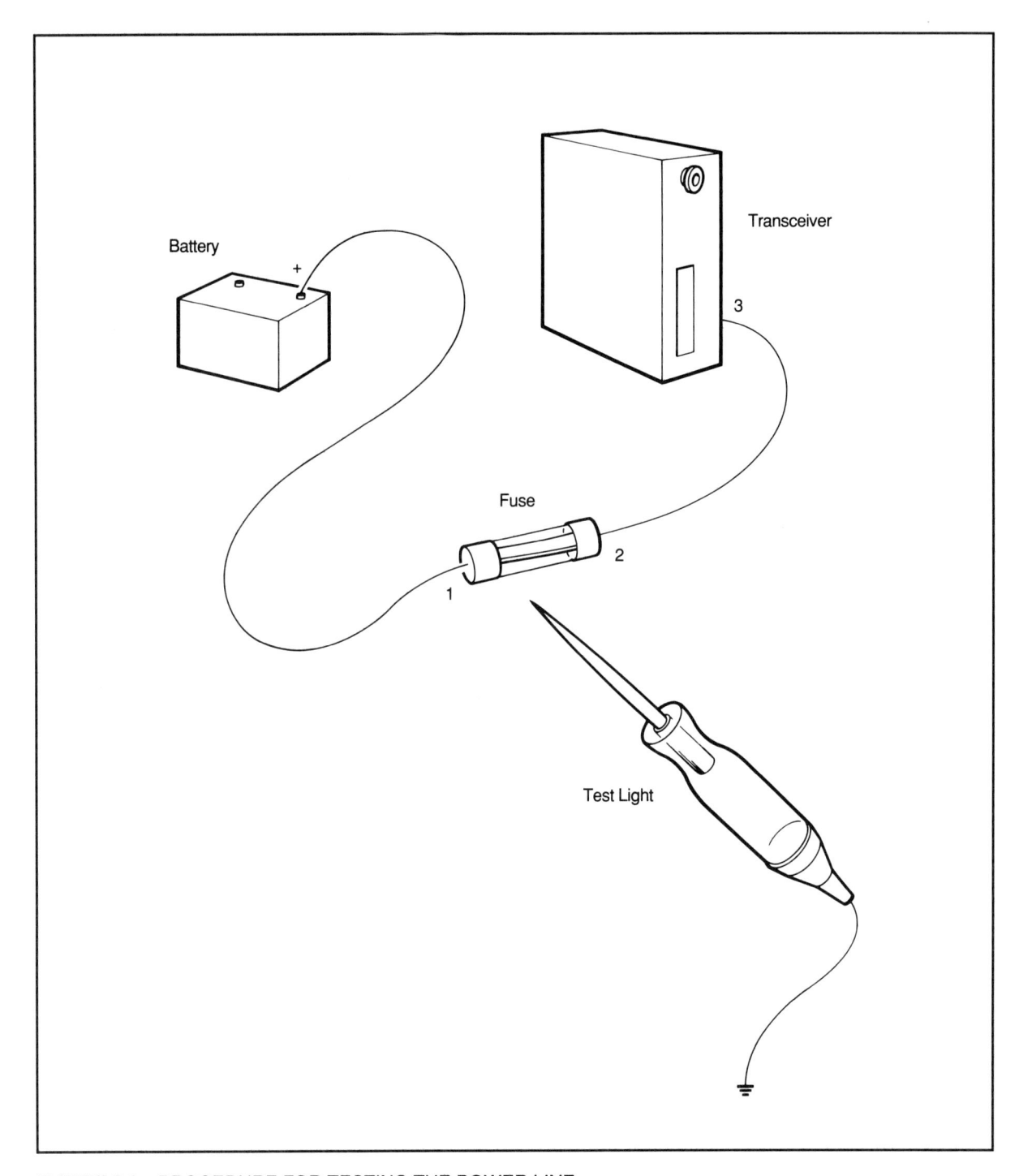

FIGURE 9.3 PROCEDURE FOR TESTING THE POWER LINE

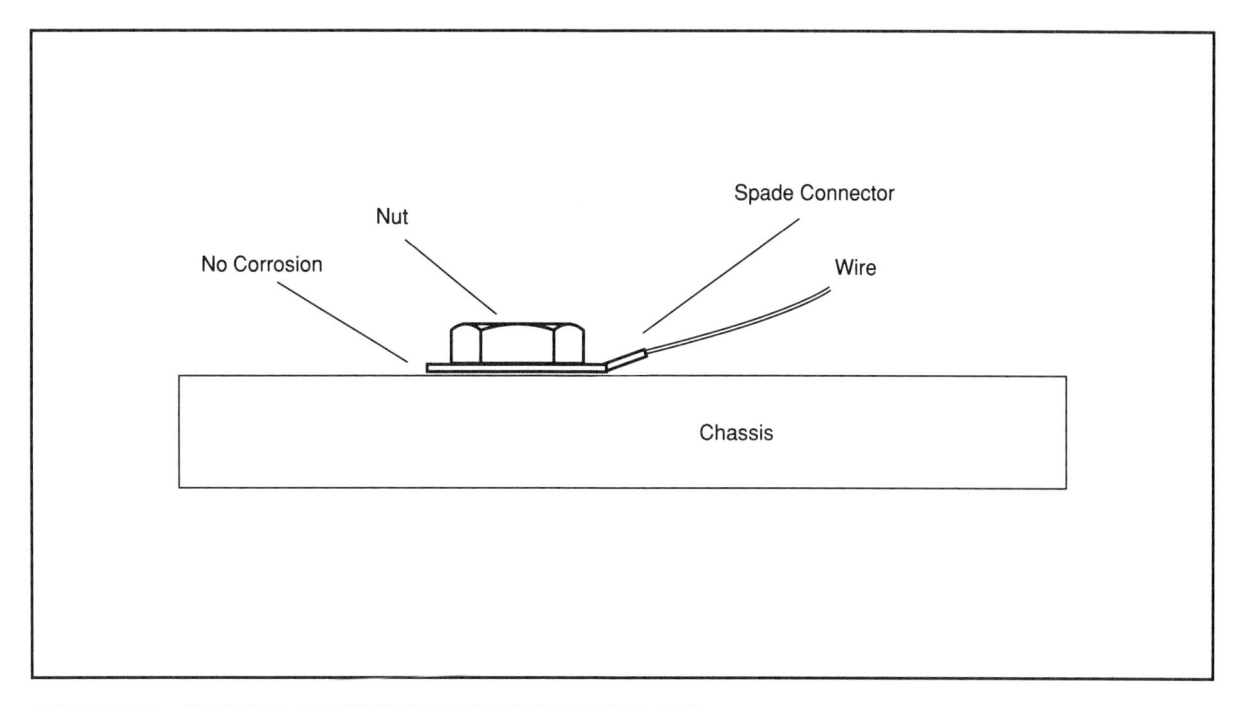

FIGURE 9.4 GROUND CONNECTION TO VEHICLE CHASSIS

appears, check the fuse on the ignition sense line. Many manufacturers fuse both the constant 12 volt and ignition sense lines. If this fuse is good, try turning on the car radio. If the car radio doesn't work, the fuse for the car radio may be blown. If this happens, the cellular phone won't turn on, since it is connected to the car radio power line after the fuse.

The next step is to check the data cable for proper operation. There is an easy way to do this—and a hard way. The hard way is to check each line in the cable to be sure that it is not shorted or broken. Checking a 10-conductor cable can take a good 30 minutes. I don't suggest doing this. The easy way is to simply disconnect the data cable and replace it with a cable that you know works. It is not necessary to actually run the cable under the carpet to do the test. Just disconnect the "suspected" data cable from the control head and transceiver and connect a working one. If the cable is run outside the car, the entire test should take only a few minutes. Turn the phone on and see if it works. If it does, remove the bad data cable and do a quick visual inspection to determine if the cable was cut by a sharp edge in the car. If it was, find a different path to route the new cable.

The last reasonable cause for a cellular phone that won't turn on is a defective circuit board. The first board to check is the power module. If this module is defective, the phone won't turn on. Before replacing it, smell the board. When these power boards malfunction, they usually burn up a few components in the process. When this happens, a rather unpleasant smell remains on the board. In this case, your nose can be a rather sophisticated piece of test equipment. If the power board is okay, try replacing the logic board.

THE PHONE DROPS TOO MANY CALLS

Another frequent complaint from customers is that their phone drops a lot of calls. A common complaint sounds like this:

My phone keeps cutting off my calls. I make a call and talk for a while. All of a sudden, I hear static and my call gets cut off. Then I want to throw the damn phone out the window. It seems like I can't make a call without this happening.

Dropped calls are probably the most common customer complaint in the cellular industry. They have a wide range of causes and can be rather difficult to track down.

The most common cause of dropped calls is improper antenna installation. About two years ago, Dr. Herschel Shosteck and I did some research on the causes of dropped calls. We came to the conclusion that improper antenna installation was the cause in the vast majority of situations. Improperly installed antennas have a tendency to be directional and thus cause the phone to undergo more handoffs than would be necessary with an omnidirectional antenna. The highest probability of a dropped call occurring is during a handoff. It then stands to reason that the more handoffs there are during a call, the greater the chance of the call being dropped. We were able to substantiate this theory with some field tests and found that an improper antenna installation could cause a cellular phone to drop three to five times as many calls as normal.

What is normal? That depends upon the system, the length of the call, the area the call is placed in, the time of day, and maybe even the phase of the moon (well, perhaps not the phase of the moon). This is another one of those situations where you should get to know some of the engineering and customer service people at the cellular service provider. They will know where the system works and where it doesn't. They will also know what a "normal" amount of dropped calls is.

When a customer complains about a lot of dropped calls, the antenna should be checked first. Remember that not all of the installations you will be repairing will have been done at your shop. Some service facilities may not take as much care in performing their installations as you do, so don't take anything for granted when looking at another shop's work.

The most common mistake made by inexperienced installers is to use the wrong antenna or to mount the antenna in the wrong place. Before making any electrical checks, look at the antenna installation. Here is a list of some common antenna installation mistakes or problems, along with corrections for them.

1. ROOF-TOP ANTENNA ON THE TRUNK. Although this antenna installation looks nice, it doesn't work. It will cause the phone to drop calls like crazy. The larger and more complex the cellular service provider's system is, the worse the problem will be. Replace the roof-top antenna with an elevated-feed antenna, and the number of dropped calls will be reduced dramatically. Your customer will love you for it.

2. NO ANTENNA. This may sound obvious, but sometimes antennas are stolen. The phone may still work in some areas because the signal will radiate out of the mounting hardware, although not very well. I had my antenna stolen and drove around for two days before I realized it. I thought I had a problem with my phone or that Cellular One in Tampa was making system adjustments. In reality, their system was good enough so that the phone actually worked (not great, but okay) without the antenna. All carriers should have systems that work so well.

3. GLASS-MOUNT ANTENNA ON DEFROSTER WIRES. This situation will really deform the radiation pattern of the antenna. This happens because the rear window defroster actually becomes a secondary antenna and radiates signal in an inefficient manner. The installer has two choices in this situation. The first is to simply replace the antenna and coupling box. The second is to cut the defroster wires. Be sure to check with the customer before cutting the wires, since you can't "uncut" them once the job has been done.

4. QUARTER-WAVE (0-DB GAIN) ANTENNAS. Generally, these antennas don't work very well. They look great, but they don't have the gain to work in marginal areas. With these antennas, the installer needs to do a bit of detective work with the customer. If most of the dropped calls occur in fringe areas, the antenna is suspect. However, if most of the dropped calls are in a well-covered urban area, the antenna is probably not at fault. Quarter-wave (0-db gain) antennas may actually perform better than 3-db gain antennas in urban areas. The solution is to replace the antenna with one that provides 3 db of gain.

5. BENT OR TWISTED ANTENNA WHIPS. This situation will create havoc with the radiation pattern of an antenna. The antenna should also be as close to vertical as possible. An antenna that does not have an even horizontal radiation pattern will wreak havoc on the performance of a phone. Replace any antenna whip that you are not sure of.

6. DIRTY CONTACTS. These reduce the amount of power delivered to the antenna. Check the antenna connectors on the transceiver and at the base of the antenna mounting plate. If they are dirty or slightly corroded, they can usually be cleaned with a wire brush. If they are severely corroded, replace them. To keep the new contacts from being damaged by the extreme environment encountered by a car, spray them with some WD-40™ or a similar silicon-based lubricant. Carefully wipe off the excess so that it won't stain the paint or carpet in the trunk.

7. GLASS-MOUNT ANTENNAS ON ROLLS-ROYCES (OR OTHER HIGH-END CARS). Glass-mount antennas won't work on these cars because of the high amount of lead in the glass. The lead is used to make the glass more transparent. Glass-mount antennas also won't work with most aftermarket glass tinting because of the metal in the tint. The solution is to use a roof-top or an elevated-feed antenna on these cars. The tint can be removed with a razor knife if the customer wishes.

If the antenna is the right type and is in the right place, inspect the antenna connector. Check to see if it is properly crimped (remember, no crimping with pliers). Using a multimeter, check if the cable is shorted. This is done be removing the antenna from the base (it should unscrew). With glass-mount antennas, unplug the cable from the coupling box. Then, as shown in figure 9.5, place one probe on the outer shielding and one on the inner conductor. With the multimeter set to ohms, there should be infinite resistance (indicating an open circuit) between the shield and the inner conductor. If you decide that you don't want to bother removing the antenna before performing the test, you will not get a reading of an open circuit. There may be a reading of as little as 50 ohms, but the meter should not indicate a direct short.

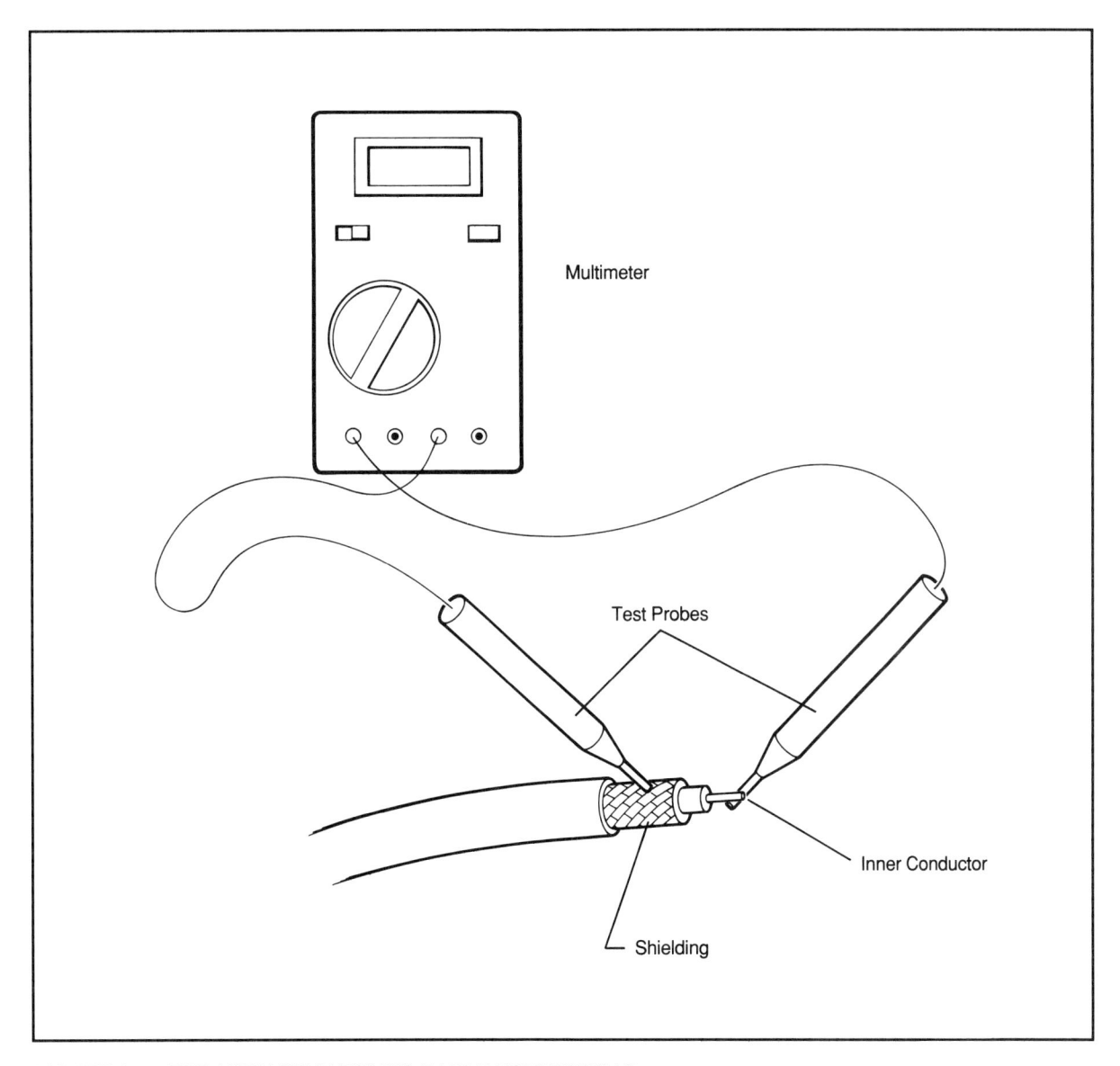

FIGURE 9.5 END VIEW OF ANTENNA CABLE FOR TESTING

184 TROUBLESHOOTING AND REPAIRING INSTALLATIONS

The cable may test okay with a meter, but remember that a meter reads DC resistance. Cellular phones operate in the 800-MHz band, and electricity has very different properties at these frequencies. An 800-MHz signal tends to flow down the outside of a cable. As shown in figure 9.6, almost all the 800-MHz signal is flowing at the outside of the conductor. A DC signal flows through almost all of the wire (see figure 9.7). The analysis for this property of high-frequency signal is rather complex and boring, but it is important because this "skin effect," as it is called, creates problems in a cable that has been "nicked" or "kinked." The electrons have trouble flowing through an imperfect edge in the cable. Look at the antenna cable to see if it has been damaged in any way during the installation process. If the cable is shorted or appears to be damaged, replace it with a new one.

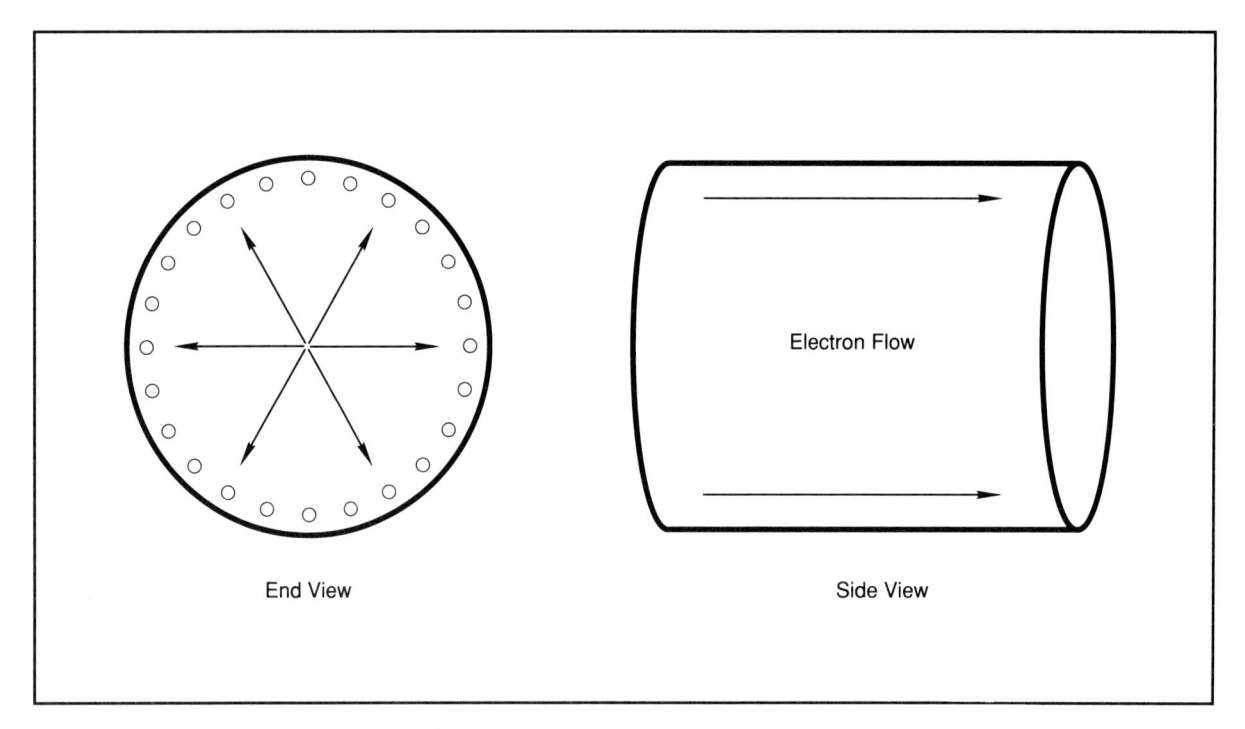

FIGURE 9.6 SKIN EFFECT AT 800 MHz
Electrons propagate near the surface of the wire.

The last test to perform on the antenna system is done with a watt meter. In an earlier chapter, we described how the watt meter is used to measure the forward and reflected power to an antenna. As shown in figure 9.8, connect the watt meter between the transceiver and the antenna. Place a test call and measure the

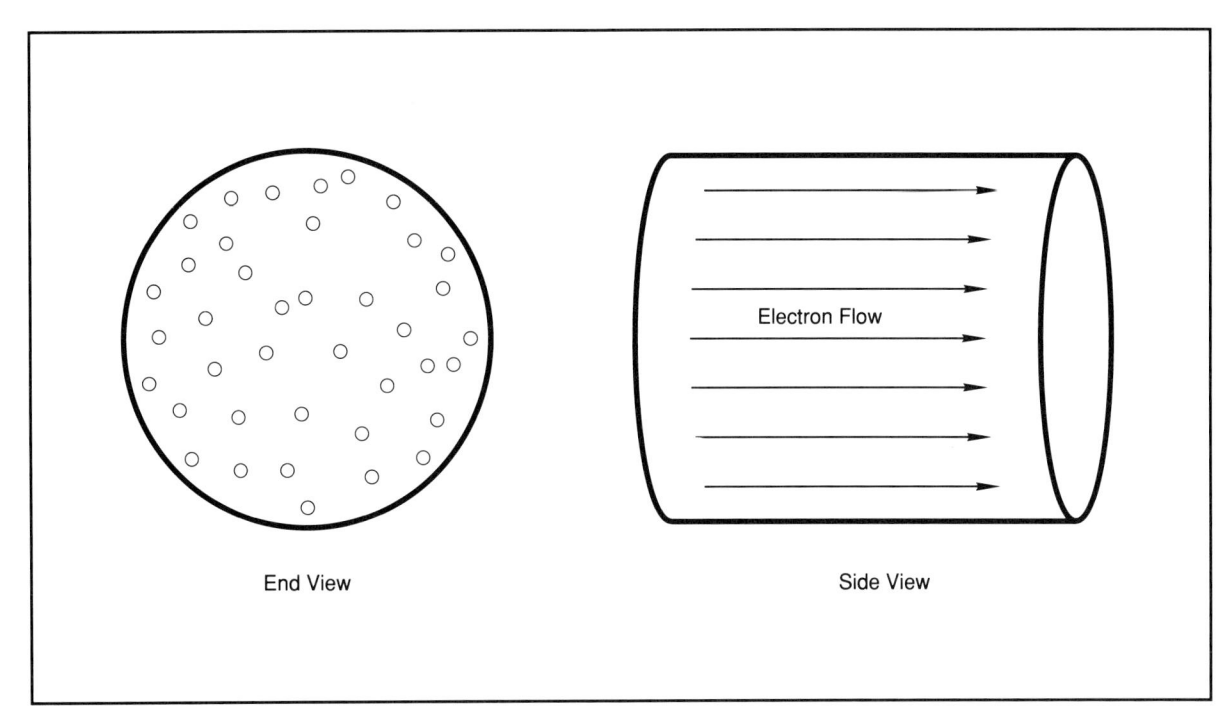

FIGURE 9.7 DC CURRENT FLOWS THROUGH THE ENTIRE WIRE

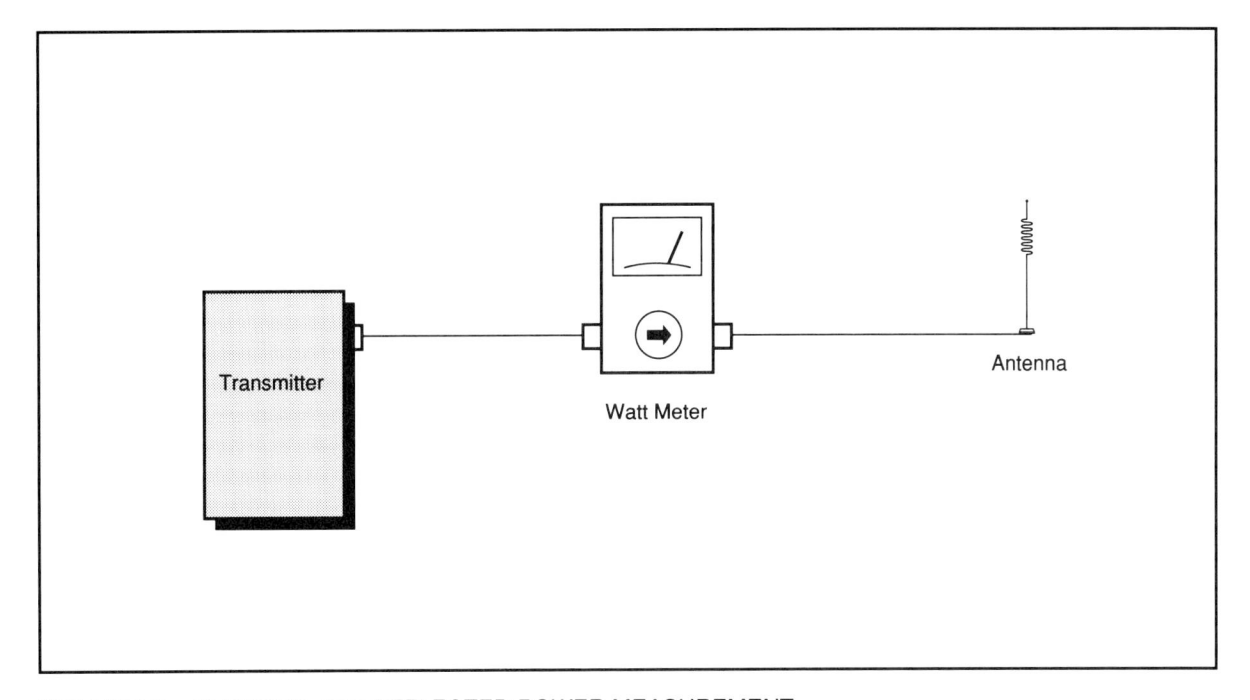

FIGURE 9.8 FORWARD AND REFLECTED POWER MEASUREMENT

186 TROUBLESHOOTING AND REPAIRING INSTALLATIONS

forward power. It should read about 3 watts. If the forward power is significantly less that 3 watts, the problem presumably lies with the transceiver. (*Note:* Some cellular service providers adjust the power of cellular telephones to reduce interference problems. If they do this in your area, the cellular telephones you test may normally operate at less that 3 watts.) If the forward power is correct, adjust the watt meter to read reflected power. This is done on most meters by rotating a knob on the front panel (see figure 9.9). A properly operating antenna should not reflect more than 0.1 watts, or 100 milliwatts. If the antenna reflects more than 0.1 watts, it is a good idea to replace the antenna.

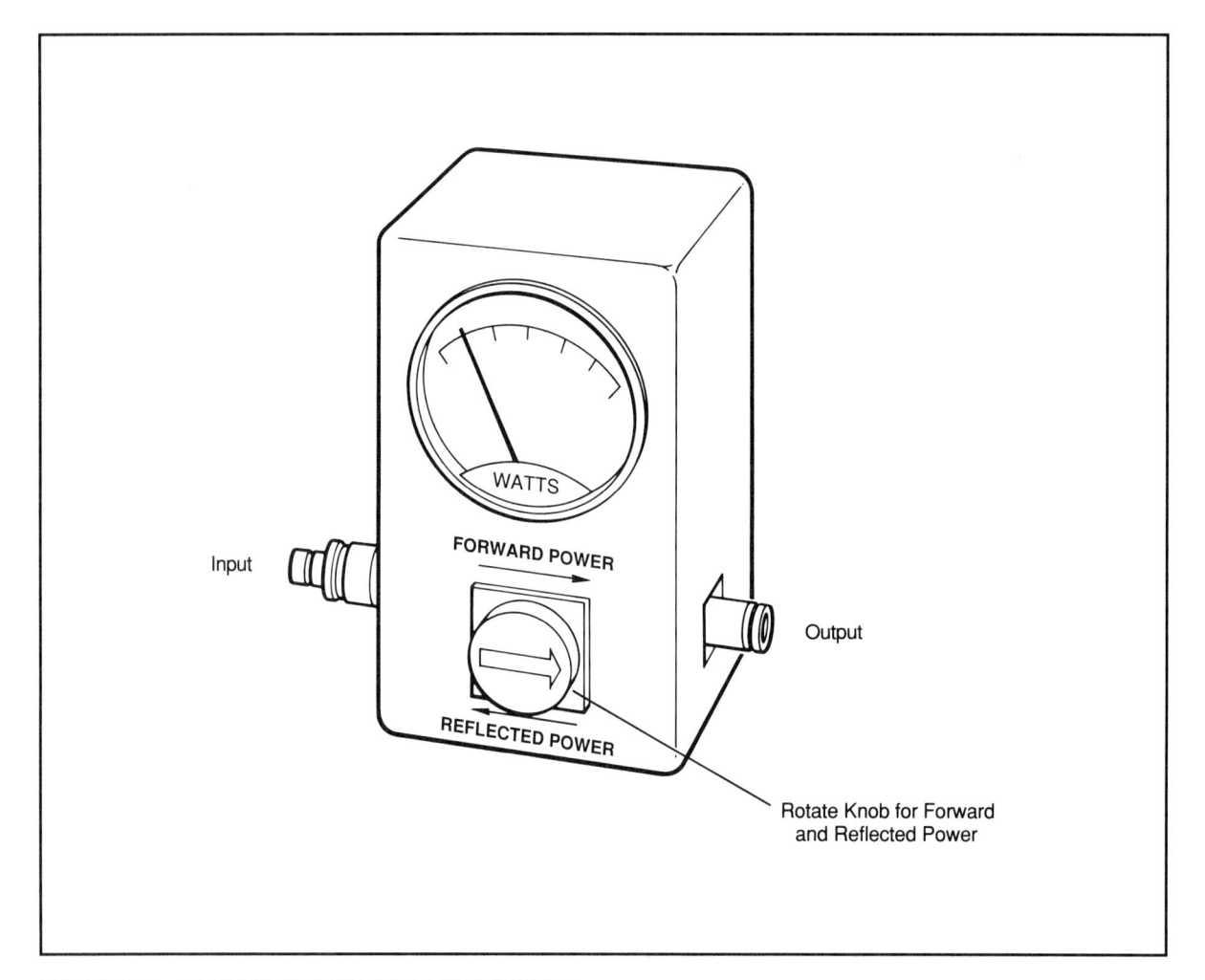

FIGURE 9.9 TYPICAL DESIGN OF A WATT METER
A watt meter is used to ensure proper operation of an antenna.

If you determine that there is no problem with the antenna system, the transceiver must be checked. Transmitter or receiver malfunctions are the most common transceiver failures to cause handoff problems. The easiest way to determine if the transceiver is at fault is to connect the cellular phone to your cellular test center. First, run the auto test. Pay special attention to any error messages that relate to faults in the transmitter or receiver.

If no problems are found with this test, switch your test center to manual mode. Reduce the RSSI to about –112 dbm, and manually force the cellular phone through some handoffs. A properly working cellular phone should easily handle a handoff at this low signal level. Run at least ten handoffs. If the unit fails to properly execute a handoff at this signal level, increase the signal level to –100 dbm. If the phone handles all the handoffs at the increased signal level, the receiver board is probably defective. Replace it, and repeat the test. If the unit still performs the same way, replace the logic board, and then the transmitter, if all else fails.

Sometimes a customer complains about dropped calls, but after extensive testing of the phone, no problems are found. This can be maddening. Take heart—this does happen. Sometimes there is nothing wrong with the phone. Almost all cellular systems have problems. Some of these problems are minor, and some will cause the phone to drop calls. This is another situation where you will need to gather information from the customer and then discuss your data with someone from the cellular carrier. Find out from the customer how often calls are dropped. Prompt them to be honest. The number of dropped calls is almost always exaggerated. Try to get the customer to give an accurate representation of the following:

1. NUMBER OF CALLS DROPPED. Accurate information will help the cellular carrier to determine if the customer is experiencing more dropped calls than is normal. After a bit of experience, you will develop a feel for what is "normal" on the carrier system. Be sure that the customer knows that a few dropped calls are normal. Most carriers give credit for dropped calls.

2. AVERAGE LENGTH OF A CALL. The longer a call lasts, the greater the chance of driving under a bridge or into a place with poor coverage. If customers tell you that their calls usually last for an hour and they almost always get cut off, it is a safe bet that

there is nothing wrong with the phone or the system. If they place two-minute calls and are dropped every time, look for a problem with the system or the phone.

3. LOCATION OF DROPPED CALLS. In a system that covers hundreds of square miles, it stands to reason that there will be a few areas where the RF coverage is not good. This is not a bad reflection on the carrier—just a reality of life. There are always going to be a few places that cannot be properly covered because of terrain, other interference sources, or zoning problems. In larger systems where channels are reused, co-channel interference can cause calls to be dropped. Co-channel interference occurs when a cellular telephone receives a signal on the same channel (RF frequency) from two different cell sites. This type of interference "confuses" the receiver and logic in the cellular phone and can cause a call to be dropped. The cellular carrier should know (and be willing to share information about) where the trouble spots are. Your approach to getting this information will determine how cooperative the carrier will be.

4. TIME OF DAY. The time of day that the dropped calls usually occur is important in larger, congested systems. During periods of high traffic volume, cells can become congested. These cells have no available channels for conversations. Either the calls are dropped or the MTSO attempts to route the conversation to a nearby cell. Sometimes this works and sometimes it doesn't, resulting in a dropped call. Once again, the carrier has statistics on cell-site loading, and your ingenuity and attitude will determine how much information the carrier is willing to share.

When this information has been completely and accurately collected, call your contact with the cellular service provider (by now, I hope you will have been convinced of the importance of doing this), and discuss the problem with them. Take a great deal of care in presenting your information calmly and clearly to the people at the carrier service. They are people just like you, and if the information is presented in an aggressive and confronting manner, they are likely to be less than cooperative with you. If you approach them with the attitude of resolving a problem with a customer that belongs to both of you, you are going to get better results. The information you give them will allow them to determine if the problem is with the cellular system and not with the subscriber's equipment.

THE PHONE NEEDS TO BE UNLOCKED

Locked phones are another problem that cellular service centers have to deal with. The customer, or someone else using the phone, accidentally locks it and doesn't remember the unlock code. If the service center keeps proper records, the solution is an easy one. Simply look up the customer's unlock code in the computer or in the paper file. Enter the code, and send the customer on his way. But what do you do if the phone was installed by another facility and you don't know the unlock code? Call them and ask for the unlock code? Not likely. You will have to rely on the equipment in your shop.

If the phone is programmed by using a PROM, the solution is easy. Remove the NAM, and insert it into the proper socket on your NAM programmer. Set the programmer to read data instead of to program. All NAM programmers have this capability. Consult your operator's manual for proper instructions. Scan the information until the unlock code appears on the screen, and copy it down. Put the NAM back into the transceiver, and enter the unlock code into the keypad. The phone should then unlock.

If you are not lucky enough to be working on a phone with a removable NAM, there are still solutions to the problem. They just require a bit more ingenuity to find. If the phone is programmed using a service handset, there is usually a command to override the lock feature or, at least, to display the lock code. If the phone is programmed from the customer handset, the problem is still manageable. Most manufacturers allow the unit to be put into a programming mode (from the customer's handset) using a special access code, even if the unlock code is not known. Once in the programming mode, the installer simply changes the unlock code and exits from the programming mode. The new unlock code can then be entered, and the phone will work normally.

If none of these solutions are applicable to your problem, call the manufacturer and ask them how to unlock a phone without the unlock code. You will not be the first person to ask this question, and they will undoubtedly have a procedure to deal with the problem.

Should you run into a situation where the manufacturer can't (or won't) help, there is one last course of action that you can take. Most cellular telephones have a small lithium battery on the logic board to keep the RAM (random-access memory) active. This battery keeps the phone from losing the numbers stored in memory when it is shut off or removed from the car. The battery has a life of several years. By desoldering and removing the battery from the circuit board, all the memory will be reset to its original state. Remove the battery, and leave it out for about an hour. Be sure to use a static discharge strap to keep from damaging the CMOS components on the circuit board. Removing the battery will give any capacitors a chance to completely discharge, and the RAM will return to the zero state. Then, resolder the battery in place, and replace the logic board in the transceiver. The customer's phone should be unlocked when it is turned on.

As a last resort, the unit can always be sent via overnight mail to the manufacturer to be unlocked. It is not the most ideal way to do the job, but it works.

One word of caution about locked phones: Some manufacturers allow only three attempts to unlock a phone. After three incorrect attempts, the phone will not allow any new attempts for about five minutes. After three more incorrect attempts, the interval goes to ten minutes, then twenty, and so on. You can see the problems that might arise after a few dozen wrong unlock codes. If you have one of these phones (check with the manufacturer), it is best to let it sit overnight before making any more attempts to unlock it.

THE PHONE TURNS ON AND OFF BY ITSELF

A rather uncommon but vexing problem is the customer who arrives at the shop with this complaint:

> *My phone turns itself on and off. I'm not touching the power switch, and it seems to do this at random intervals.*

The first thing you should do is look for loose connections in the installation. Next, check for loose circuit boards in the phone. If no problems are found in the wiring or the phone, there is one more problem to look for.

Most cellular phones have under and over voltage protection. This guards the phone's internal circuit from being damaged by too much or too little voltage. Most phones will operate on anywhere from 9 to 15 volts, although this range varies among manufacturers. If the vehicle supplies power outside of that voltage range, the cellular phone will shut itself off to prevent damage. Automobiles are equipped with devices to regulate the DC power being supplied to the electrical equipment in the car. This device is called a *voltage regulator* (what a surprise). If the vehicle's voltage regulator is defective, the voltage of the car will no longer remain at a constant 12 volts. The voltage will vary from about 10 to 20 volts, depending on engine RPM and which other appliances are being used at the time. This will cause the phone to turn on and off as the vehicle voltage varies.

I once heard a customer complain that his phone would turn off every time he shut off his vehicle air conditioning. We spent the better part of a day trying to figure out what we had done to the vehicle's air conditioning system. The fault turned out to be a bad voltage regulator. When the air conditioning compressor was shut off, the load on the vehicle electrical system would be decreased, and the voltage in the car would jump up to almost 20 volts. Consequently, the phone would shut down.

The easiest way to check for this problem is to place the probes of your multimeter across the car battery. Be sure the meter is set to read DC volts, not ohms. If the meter is set to ohms, it will be seriously damaged by the battery. When the engine is idling, the voltage across the battery should be about 12 or 13 volts. Have someone "rev" the engine. The voltage across the battery should not vary more than about 1 volt when the engine is "revved." If the voltage swings wildly, tell the customer he/she has a bad voltage regulator and that it should be replaced.

FAILURE DUE TO AGE AND USE

Older cellular phones are subject to mechanical failures that occur due to age and use. An older phone exhibiting any problems should be checked for worn handset cords, broken hand up clips, etc. A cellular phone with a worn out handset cord can exhibit problems ranging from being completely inoperative to bad audio. A good visual inspection will determine if the cord is worn out.

Cars are driven in both good and bad weather. Unfortunately, no one builds a waterproof cellular phone. Occasionally, a customer will bring a phone into the shop that appears to be undamaged, but it just doesn't work. A good installer will be on the watch for water damage in a cellular phone. Once a phone is water-damaged, the warranty is usually voided, and the entire phone must be replaced. There are a couple of ways to detect water damage, and it is in the installer's best interest to learn them. If you incorrectly diagnose a water-damage problem and return the phone for warranty repair, you will find that you are not covered for the problem. If the customer has left with a new phone or circuit board, the service facility is "out of luck."

There are two quick ways to tell if a phone has been damaged by water. The easiest is to look at a paper label on the transceiver. Almost all manufacturers have them on the transceiver somewhere. If the phone has been wet, the ink on the label will be smeared. If no paper label is present, look closely at the circuit board. A water-damaged circuit board will almost always have some residue on it. With a quick look, you should be able to figure out if the phone has been wet.

THE HANDS-FREE UNIT SOUNDS TERRIBLE

The last part of the phone that can cause problems is the hands-free circuit. The following is a common complaint from a customer:

My hands-free sounds terrible. The people I call tell me I sound like I am calling from a garbage can.

This problem can result from a number of causes. The most common is misrepresentation or inadequate explanation of the device's capability. All hands-free units sound hollow. The person who invents one that sounds as good as the audio from a handset will make a lot of money. When customers buy a phone, you should explain to them that all hands-free units sound like this. If they don't believe you, demonstrate a few different brands to them. When this is done, they leave with reasonable expectations of the equipment they have just purchased. Let them know how it works and how it will sound, and they will be a lot happier.

Hands-free units do fail at times. The cause is almost always the microphone unit. Simply replace it, and the customer can be on his/her way in a few minutes. Before removing the suspected unit, plug a working hands-free microphone into the hands-free plug on the control head. Try to place a call using the new microphone before replacing the one in the vehicle. This will save you the trouble of replacing a working microphone. If the microphone needs to be replaced, be sure to recalibrate the hands-free system.

Another reason for a hands-free system not working is an incorrectly programmed phone. Every NAM (regardless of the type of NAM being used) has a bit for hands-free operation. If the bit is incorrectly set, the hands-free unit will not operate. This bit won't change after the phone has been programmed, so if the hands-free unit stopped working after a month, don't bother to check the NAM. If the phone never worked, the NAM is a good place to start.

Poor sounding hands-free systems represent a larger challenge for the installer. A poor sounding hands-free unit is usually the result of poor installation or placement of the microphone. As discussed in the installation section, the hands-free microphone picks up all of the sound in the car, not just the driver's voice. The best place for the microphone is on the sun visor or on the headliner trim on the roof. Stay away from "fancy" custom mounting jobs on the dashboard, steering wheel, or anywhere else. Another cause of a poor sounding hands-free unit is improper calibration. If you can't find anything wrong with the hands-free system, try recalibrating it.

REPAIRING A HOLE IN THE GAS TANK

Last but not least in the troubleshooting section, you learn how to fix your own mistakes. If you do enough installations, you are going to drill a hole in a gas tank at some point. When this happens, it is very embarrassing. Fortunately, the hole can be repaired. If (or I should say when) this happens, the first thing you should do is let the customer know. Taking the "up front" approach with customers is your best bet. If they find out at a later date that their gas tank was damaged during the installation, they will be furious and may even bring legal action against the service facility. Admitting to the error before they are aware of it will usually diffuse the customer's anger.

There are two approaches to fixing a gas tank. The best approach for the installation facility is to repair the tank. With a little ingenuity, the tank can be fixed quickly and inexpensively. If the hole is small, drive a screw into the hole that is slightly larger than the hole (see figure 9.10). This will seal the largest

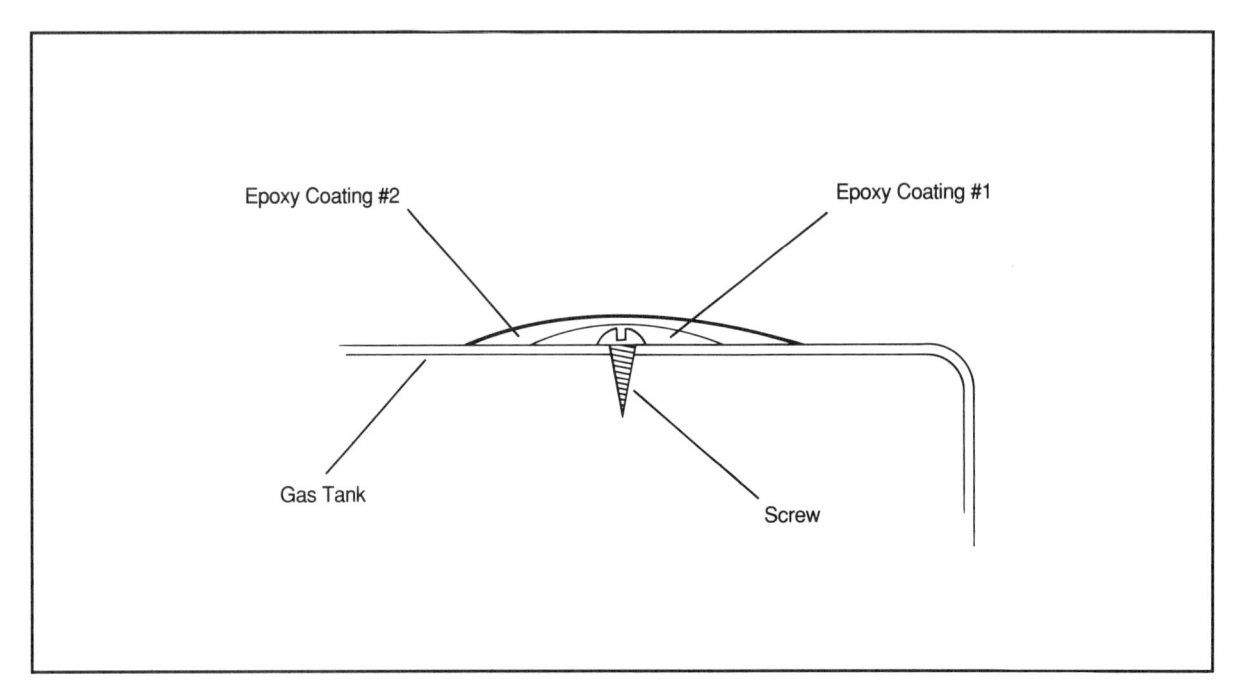

FIGURE 9.10 "QUICK" REPAIR FOR A GAS TANK

part of the hole. The rest of the hole can be sealed with gas-resistant epoxy (available at most hardware stores). Mix the epoxy, and apply it around the base of the screw after completely cleaning the area. The area must be cleaned, or the epoxy will not stick to the gas tank. After the epoxy has dried, apply a second layer to ensure a complete seal.

A word of caution about this procedure: First, be sure that the epoxy used is the type intended for use in environments where gasoline is present. Gasoline dissolves many adhesives, so be sure to use the right kind of epoxy. The second caution is to be sure NO ONE is smoking while this work is being done. A hole in the tank will allow gasoline vapors to escape. It is these vapors that are explosive, not the liquid gasoline. It doesn't take a lot of vapor to create a very explosive gasoline/air mixture. Be sure that adequate ventilation is present and that no sparks, cigarettes, etc., are present.

If the customer is not willing to have the gas tank repaired, don't despair. Gasoline tanks are not very expensive or difficult to replace. A tank can usually be purchased from a dealer and installed without too much difficulty. If you don't want to perform this task, have the dealer do it. In most cases, it shouldn't cost more than about $250 for the entire job.

REPAIRING A HOLE IN THE FUEL LINE

Holes drilled in fuel lines can also be repaired easily. If a section of the fuel line is accidentally drilled, simply replace it with a short section of rubber (gasoline-resistant) fuel line. Rubber fuel line is quite inexpensive; it is a good idea to keep a few feet of it on hand. To replace a section of fuel line, cut the damaged section out with either a pipe cutter or a hack saw. Be sure to have something to plug the line with, since fuel will drain from the gas tank after the line is cut. Slide the new rubber fuel line over the undamaged metal line, and secure it with small hose clamps. Seal the junction with some gasoline-resistant epoxy to prevent any fuel leakage (see figure 9.11). The entire repair should not cost more than a few dollars, and the customer should be happy with the repair. Remember, it is important to be "straight" with the customer about any repairs that have to be made to the car.

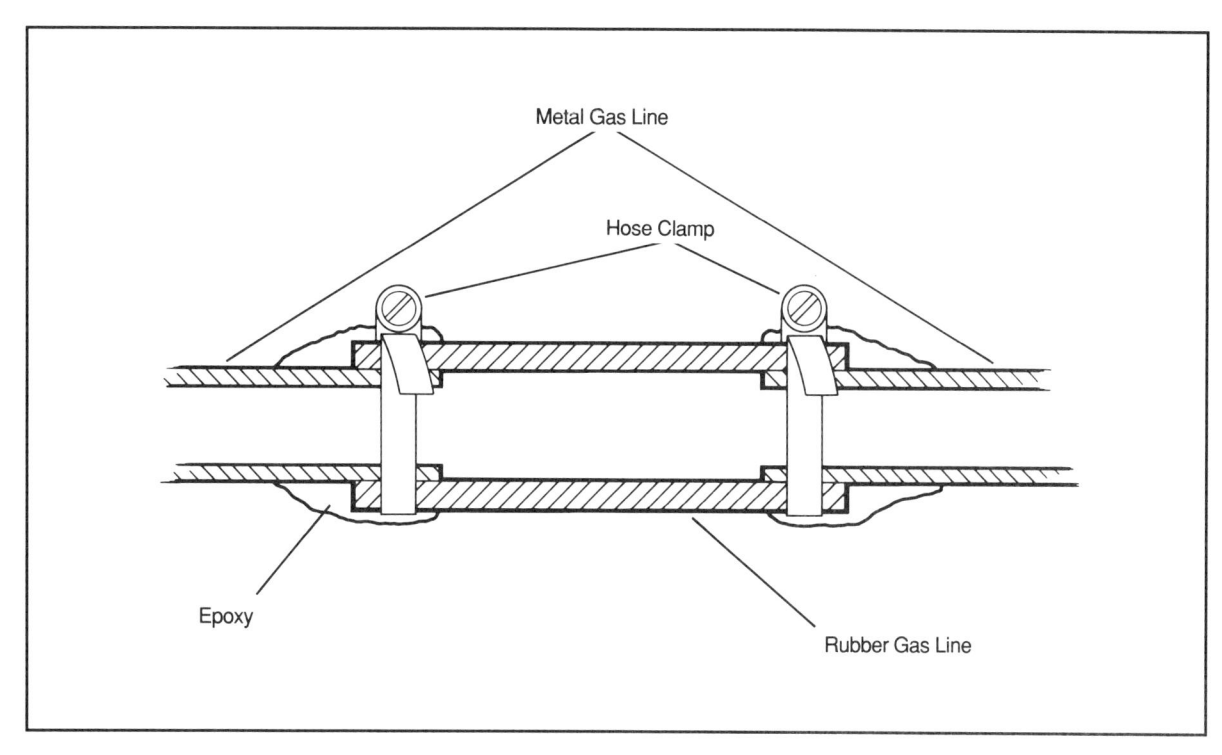

FIGURE 9.11 "QUICK" REPAIR FOR A FUEL LINE

SUMMARY OF TROUBLESHOOTING AND REPAIR PROBLEMS

1. **The phone turns on, but it has never been able to place or receive calls.**

    a. Using the autotest mode, check all the functions of the cellular phone. If a fault is found, replace the defective module, or return the transceiver to the manufacturer for repair. If no fault is found, refer to problem 2.

    b. If you don't have a cellular test center handy, try to place a call on the alternate carrier's system. If the call processes properly, refer to problem 2. If the phone will not process a call, either troubleshoot it with a cellular test center or return it to the manufacturer for repair.

2. **The phone turns on and shows no fault on a cellular test center, but it has never been able to place or receive calls.**

    a. Call the cellular service provider, and verify that the electronic serial number and telephone number are programmed into the phone properly. If they are, check the carrier's programming to be sure they have the correct ESN and phone number.

    b. Have the carrier check to see if the customer's ESN has been placed into the negative directory.

3. **The phone used to be able to place calls, but now it can't. The phone will process calls on a cellular test center.**

    a. Be sure the customer has paid his/her bill to the cellular carrier.

4. **The phone will place calls in some areas but not in others.**

    a. The customer is probably trying to place a call in a fringe area or an area with poor coverage. Locate the areas on a coverage map to be sure.

5. **The power output level from the transceiver is low.**

   a. Referring to the manufacturer's service manual, adjust the power output to a full 3 watts. Don't adjust the power level any higher than 3 watts. It will not result in any better operation, and it will shorten the life of the phone. Besides, it's illegal. ✓

   b. Replace the transmitter board in the phone. ✓

   c. Replace the logic board in the phone. ✓

   d. Return the phone to the manufacturer for repair. ✓

6. **The phone will not turn on.**

   a. Make sure the power switch on the control head is turned on. ✓

   b. Check for blown or missing fuses. ✓

   c. Be sure that the power cables are securely connected to the vehicle battery, ground, and the transceiver. ✓

   d. Be sure the data cable is properly connected to the transceiver and the control head. ✓

   e. Verify that the data cable is not defective. ✓

   f. Replace the power board in the transceiver. ✓

   g. Replace the logic board in the transceiver. ✓

   h. Replace the control head. ✓

7. **The phone drops too many calls.**

   a. Check for an improper or defective antenna installation. ✓

   b. Check for a defective antenna connector. ✓

   c. Replace the receiver board in the transceiver. ✓

   d. Replace the logic board in the transceiver. ✓

198 TROUBLESHOOTING AND REPAIRING INSTALLATIONS

    e. Replace the transmitter board in the transceiver.

    f. Poor coverage or a congested system can cause dropped calls. Call your carrier to see if they have problems in the areas the customer is complaining about.

8. The phone needs to be unlocked.

    a. Look up the customer's unlock code in your computer or files.

    b. If the phone uses a NAM, remove it, and read the contents with a NAM programmer.

    c. If the phone is programmed using a service handset, enter the programming mode, and read the unlock code from the phone.

    d. Call the manufacturer, and ask their service department for help (yes, I am serious).

    e. Remove the lithium backup battery from the logic board, and let the unit "sit" for a couple of hours. Attempt this only if all else fails because there is always the chance that you will damage the board in the process.

9. The phone turns on and off by itself.

    a. Check for loose connections.

    b. Check for a defective voltage regulator in the car's electrical system.

10. The phone is water-damaged.

    a. Try to clean the circuit boards with a fluorocarbon-based cleaner.

    b. Throw the phone out, and sell the customer a new one. Manufacturers' warranties don't cover abuse (which is what water damage is), and they will not repair the unit.

11. **The hands-free unit sounds terrible or does not work at all.**

   a. Replace the microphone. ✓

   b. Move the microphone to a place where it is closer to the driver's face. ✓

   c. Check the cellular telephone's programming to be sure the hands-free bit is set to a value of "1." ✓

APPENDIX A

Glossary

A/B switch A feature in a cellular telephone that allows the user to select the carrier that the phone will use when it is first turned on.

AC power Alternating Current power. The type of power available from the wall sockets in your home. Standard power available in a home in the United States is 117 volts AC at 60 Hz. In some parts of Europe, it is 220 volts AC at 50 Hz.

Access The process of connecting a cellular system via a control channel.

Access channels A control channel used by a mobile terminal to access a cellular system to obtain service.

Amperes A measurement of current, or the number of electrons flowing through a cable per unit time.

Amplitude The strength of a signal.

AMPS Advanced Mobile Telephone System. The cellular communications system used in the United States, Canada, Australia, and a number of other countries. The specification was developed by Bell Laboratories.

Battery backup The DC power supply that connects to the volatile memory circuit of a cellular telephone. This permits data to be retained in memory even after external power has been removed.

Baud rate The maximum rate at which a data line will accept data.

BCH code Bose-Chaudhuri-Hocquenghem code. An error-correcting code used in cellular communications.

Bit The smallest piece of digital information: a 1 or a 0. A single bit can represent only ON (1) or OFF (0).

Bit field A string of bits.

Blank and burst The process of blanking the voice portion of a cellular channel and sending the data in a rapid burst, usually 100 to 350 ms.

Busy-idle bits The portion of a data stream transmitted by a cell site on a forward control channel that is used to indicate the current busy-idle status of the corresponding reverse control channel.

Byte Generally, 8 bits long. One half of a byte is called a nybble (yes, I am serious).

Call supervision The process of the MTSO monitoring call quality and the action taken based on prevailing conditions.

Carrier-to-interference ratio The ratio of the signal on a cell's channel to the signal that interferes with it from the same frequency on another cell's channel.

Cellular system A mobile telephone communications system that uses low-powered transmitters to communicate with remote (mobile or fixed) stations. These low-powered transmitters allow the reuse of channels for a very spectrum-efficient communications. The term applies not only to the domestic AMPS system but also to several systems used in other countries. *See* TACS and NMT.

CGSA Cellular Geographic Service Area.

Channel collision The process of two cellular terminals trying to use the same channel at the same moment. In cellular telephony, safeguards exist to inhibit this phenomenon.

Companding A noise reduction process used in cellular telephony. The signal is compressed before transmission and expanded in the cell site.

Control channel A channel used to transmit digital information to and from the cellular terminal.

Current/phase relationship The phase relationship of the current to the voltage in a circuit. In some circuits, the current leads the voltage; in others, the voltage leads the current.

Data bus A system that transfers data from one place to another.

Data stream Transmission of information in a digital format.

DBX™ A double-ended noise reduction system commonly found in home entertainment electronics. A professional version is used in recording studios.

DC power Direct Current power. The type of power available from a car battery. A car battery usually supplies 12 volts DC.

Dielectric An insulating material that separates the plates of a capacitor. The material can be any number of materials, such as air, plastic, mica, or ceramic. The dielectric material must be an insulator.

Digit outpulsing After receiving a cellular terminal's Mobile Identification Number and Electronic Serial Number, the dialed digits are outpulsed, or sent, to the POTS network.

Digital color code A digital signal transmitted by a cell site on a forward control channel that is used to detect capture of a cell site by an interfering cellular terminal.

Digital modulation Describes any number of modulation techniques that take advantage of modern digitization techniques.

Directed retry A software routine used to route traffic to alternate cell sites when high traffic conditions make handoff to the preferred cell site impossible. It is a tool that can increase the traffic-handling capacity of a cellular system.

Disconnect The process of ending a call, either intentionally or unintentionally.

Dolby™ A double-ended noise reduction system commonly used in home entertainment products. A professional version is used in recording studios.

DTMF Dual-Tone Multi-Frequency. The Touch-Tones™ that are used to operate answering machines, voice mailboxes, and some computer equipment.

Dynamic power-level control The process by which the MTSO adjusts the power output of a cellular telephone. This is done to reduce co-channel interference.

EEPROM Electronically Erasable Programmable Read-Only Memory. Used as a NAM in keypad-programmable cellular telephones.

Electronic lock Prevents unauthorized use of a cellular terminal. The lock code is programmed into the NAM during the installation of a cellular terminal.

Electrons Electrons flowing through a wire are electricity.

END key The key on a cellular terminal that terminates a conversation.

Erlang A unit of traffic measurement.

ERP Effective Radiated Power. The amount of power radiated from an antenna relative to a 0-db gain radiator.

Flag A single bit that can be set to a value of either 0 or 1. If the bit is set to 1, the flag is said to be set to the ON condition.

Flash request A request to the MTSO to alert the system that a cellular subscriber wishes to use a vertical service such as call waiting.

FM carrier The signal that is modulated by an audio signal.

Forward control channel A control channel from a cell site to a cellular terminal.

Forward voice channel A voice channel from a cell site to a cellular terminal.

Frequency The number of cycles per second of a signal (either audio or radio). Frequency is expressed in hertz or cycles per second.

FSK Frequency Shift Keying. A modulation technique used in telecommunications.

Gain The process of increasing the strength of a signal. Turning the volume up on your television set increases the gain of the audio portion of the television program.

Group identification A subset of the most significant bits of the system identification (SID) that is used to identify a group of cellular subscribers.

Handoff The act of transferring a cellular terminal from one voice channel to another.

Home mobile terminal A mobile terminal that operates on a cellular system from which service is subscribed.

Impedance Similar to resistance, but it describes the effect in circuits with alternating current.

IMTS Improved Mobile Telephone Service. The high-power automatic mobile telephone system that preceded AMPS in the United States and Canada. Mobile subscribers were able to directly dial calls without operator assistance, but waiting times of up to 30 minutes for service were common in large cities. Still in use in some areas.

Intersyllabic noise The noise that is heard when the speaker pauses between words.

Landline Communications that are transported using standard fixed telephone lines. These lines can be copper cable, fiber optical cable, or any similar nonradio links.

Large-scale integration An integrated circuit with a large number of transistors packed into a very small space.

Lead-acid battery A common type of rechargeable battery used in transportable cellular telephones.

Manchester code A digital encoding format used in cellular telephony.

Megahertz The frequency of a signal, in millions of cycles (or hertz) per second.

Microprocessor The computer chip that controls the cellular telephone.

MIN Mobile Identification Number. The 34-bit number that is a digital representation of the 10-digit telephone number assigned to a mobile terminal.

Mobile station class The mobile terminal classes are defined as follows: Class I = 3.0 watts; Class II = 1.2 watts; Class III = 0.6 watts.

Modulator An electronic device that superimposes a signal on a carrier wave.

MSA Mobile Service Area.

MTS Mobile Telephone Service. The manual mobile telephone system that preceded IMTS in the United States and Canada. Users of this service were required to contact a mobile operator who would dial the number of the called party. Still in use in some areas.

MTSO Mobile Telephone Switching Office. The central switching, processing, and control center for a cellular system.

NAM Number Assignment Module. The device that stores the customer-specific information required when programming a cellular telephone.

Nickel-cadmium battery A common type of rechargeable battery used in transportable and hand-held portable cellular telephones.

NMT Nordic Mobile Telephone system. The cellular communications system used in Nordic countries (Denmark, Norway, Finland, Sweden, Iceland). One of the first operating cellular systems in the world.

Noise The unwanted portion of a signal. Noise usually manifests itself as a "hiss" or "crackle." Because of its random nature, it is very difficult to filter or remove from a signal.

Nonwireline carrier A carrier that operates on the "A" band of channels. This carrier was usually formed by an independent (non-Bell) company or group of companies. In recent years, some of these carriers have been purchased by Bell Companies or other large groups with interests in telecommunications. The distinction between wireline and nonwireline companies has become blurred in recent years.

Ohm A unit of resistance. The more ohms, the higher the resistance.

Outpulse The process of sending telephone digits from one location to another.

Paging The act of seeking a cellular terminal when an incoming call has been placed to it.

Paging/access stream The data that flows between the cell site and the cellular terminal during the paging/access process.

Paging channel A forward control channel that is used to page cellular terminals and to send commands.

PCM Pulse Code Modulation.

Polarity The charge on a cable or battery terminal. The polarity of a cable or battery terminal can be positive or negative.

POTS An old telephone company term meaning Plain Old Telephone Service.

Processor A "slang" term for microprocessor, which is the computer chip that controls the cellular telephone.

PROM Programmable Read-Only Memory. Used in some cellular telephones as a NAM.

Propagate How radio waves move through space.

PTSN Public Telephone Switching Network. The switching network used by local and long-distance landline telephone carriers.

RAM Random Access Memory. The memory in a computer where the program is handled. All data in RAM is lost (usually) when the machine is turned off.

Receiver The portion of a cellular telephone that converts the radio signal to audio material so that it can be understood by the user of the phone.

Redundant Two complete systems available at one time. If the online system fails, the backup will take over with no loss of service. Most MTSOs have redundant critical systems.

Registration The process by which a cellular terminal identifies itself to a cell site as being active in the system at the time the message is sent to the cell site.

Release request A message sent from a cellular terminal to a cell site indicating that the user desires to terminate a call.

Resistance The part of a circuit that absorbs energy. Voltage across a resistance = current × resistance ($V = IR$).

Reverse control channel A control channel from a cellular terminal to a cell site.

Reverse voice channel A voice channel from a cellular terminal to a cell site.

RF Radio Frequency.

RMS Root Mean Square. Generally, average power of a signal, expressed in watts.

Roamer A cellular terminal operating on a cellular system other than the one from which service is subscribed.

ROM Read-Only Memory. A device often used to store programming information in a cellular telephone. When used for this purpose, it is often called a NAM.

RSSI Relative Signal-Strength Indicator. Indicates the strength of a cellular signal.

SAE The North American measurement system of inches, gallons, pounds, etc.

SAT Supervisory Audio Tone. One of three tones (5970, 6000, 6030 Hz) transmitted out of band and returned to the base station via the cellular terminal. SAT functions in a manner similar to DC current loop in a POTS telephone.

SIDH The system identification field in a NAM. For example, the SIDH of the New York City wireline is 00022. The nonwireline SIDH in Buffalo, New York, is 00003.

Signal-strength indicator Indicates the level of radio signal available at any particular time.

Space-diversity receive antennas Two antennas that are used to increase the sensitivity of a receiver. A special receiver has a voting circuit that selects the antenna with the strongest signal.

Spectrum The number of channels available. A lot of spectrum means that there are a lot of channels accessible to the user.

Switch A "slang" term for Mobile Telephone Switching Office.

Syllabic amplitude companding A companding scheme used in cellular telephony.

T1 A digital type of telephone protocol that is used to carry many conversations at one time.

TACS Total Access Communications System. The cellular communications system used in Great Britain.

TDM Time Division Multiplexing. A digital communications scheme used in some switching systems.

3M connector A special-purpose connector that provides a secure connection between two wires.

Transceiver An electrical device that incorporates both a transmitter and a receiver.

Transmitter An electrical device that is used to convert audio, video, or data to radio waves at relatively high frequencies.

Variable-gain linear amplifier An amplifier whose gain can be controlled externally. In a cellular telephone, this device is used in companding circuits.

Voice channel A channel on which a voice conversation occurs and on which brief digital messages from a cell site to a cellular terminal (or vice versa) can be sent.

Voltage An electromotive force.

Watt A measure of power. Watts = voltage × amperes.

Wireline carrier A carrier that operates on the "B" band of channels. This carrier has traditionally been associated with a "Regional Bell Operating Company" that was formed by the divestiture of AT&T. The distinction between wireline and nonwireline companies has become blurred in recent years.

Yagi antenna A special-purpose antenna that provides the user with a great deal of gain (typically, 9 db or more) in one particular direction.

APPENDIX B

Directory of Manufacturers

Manufacturer	Equipment or Service

Addison Group
4 Northridge Center
Pittsburgh, PA 15212
412-321-0400 (tel)

Consulting

Alexander & Associates
5223 South Street
Nacogdoches, TX 75961
409-569-0556 (tel)

Consulting

Allgon Antenna, Inc.
2100 North Highway 360, Suite 1002
Grand Prairie, TX 75050
214-641-3887 (tel)
214-641-1945 (fax)

Cellular Antennas

Alpine Electronics
Box 2859
915 Gramercy Place
Torrance, CA 90509
213-326-8000 (tel)
800-421-2284 (toll-free)
213-533-0369 (fax)

Cellular Terminals
Vehicular Antennas

DIRECTORY OF MANUFACTURERS

Manufacturer	**Equipment or Service**
Ambassador Communications 2237 South Huron Santa Ana, CA 92704 714-546-1300 (tel)	Cellular Accessories
The Antenna Company 2415 Braga Drive Broadview, IL 60153 312-450-0330 (tel) 312-739-4953 (fax)	Cellular Antennas
Antenna Electronics Box 14728 Fort Worth, TX 76117 817-281-8403 (tel)	Vehicular Antennas
The Antenna Specialists Company 30500 Bruce Industrial Parkway Cleveland, OH 44139-3996 216-349-8400 (tel) 216-349-8683 (fax)	Base Station Antennas Portable Antennas Vehicular Antennas Mounting Hardware
Anxiter Brothers, Inc. 4711 Golf Road Skokie, IL 60076 312-677-2600 (tel)	Vehicular Antennas
ARA Manufacturing Box 53400 606 Fountain Parkway Grand Prairie, TX 75050 214-647-4111 (tel) 214-641-8787 (fax)	Cellular Terminals Vehicular Antennas Mounting Hardware
Astronet Corporation 400 Rinehart Road Lake Mary, FL 32746 305-849-4600 (tel) 305-849-4920 (fax)	Cell-Site Equipment Cellular Switching Equipment
Audiovox Corporation 150 Marcus Boulevard Hauppauge, NY 11788 516-231-7750 (tel)	Cellular Terminals

Manufacturer	Equipment or Service
Autotenna, Inc. 2575 Presidio Street Long Beach, CA 90810 213-632-5555 (tel)	Vehicular Antennas
Bank of Illinois Company Box 128 Champaign, IL 61820 217-351-6568 (tel)	Billing Systems
Basel Marketing 3 Malaga Cove Plaza, Suite 207 Palos Verdes Estates, CA 90274 213-375-0559 (tel)	Cellular Accessories
Battery Pack, Inc. 25700 I-45 North, Building 111 Spring, TX 77380 713-367-9393 (tel) 800-245-1138 (toll-free) 713-292-7139 (fax)	Batteries
Bird Automotive 123 Nash Island Road Darien, CT 06820 203-656-0630 (tel)	Cellular Accessories
Bird Electronic Corporation 30303 Aurora Road Cleveland, OH 44139 216-248-1200 (tel)	Test Equipment
Blaupunkt, Division of Robert Bosch 2800 South 25th Avenue Broadview, IL 60153 312-865-5200 (tel)	Cellular Terminals
BOBCO Benjamin Fox Pavilion, Suite 823 Jenkintown, PA 19046 215-886-1470 (tel)	Consulting

Manufacturer

Bogner Broadcast Equipment Corporation
603 Cantiague Rock Road
Westbury, NY 11590
516-997-7800 (tel)
516-997-7721 (fax)

Briefcase One
Division of Trac Communications
Box 43
Madison Heights, MI 48071
313-680-1200 (tel)

Bytek Corporation
Instruments Systems Division
1201 South Rogers Circle
Boca Raton, FL 33431
305-994-3520 (tel)

C-TEC Corporation
Box 3000
46 Public Square
Wilkes-Barre, PA 18703-3000
717-825-1113 (tel)

Cartwright Communications Company
5299 West Washington Boulevard
Los Angeles, CA 90016
213-936-0073 (tel)
800-634-3948 (toll-free)
213-934-0892 (fax)

Cellsmart, Inc.
3100 Skokie Highway
Highland Park, IL 60035
312-819-2500 (tel)

Celltech, Inc.
17150 Butte Creek, Suite 200
Houston, TX 77090
713-586-8031 (tel)

Equipment or Service

Base Station Antennas

Cellular Accessories

NAM Programmers

Billing Systems

Wholesaler
Accessories

Cellular Accessories

Billing Systems

Manufacturer	Equipment or Service
Cellular Accessories 587-F North Ventu Park Road, Suite 430 Newbury Park, CA 91320 805-499-9908 (tel)	Control Head Covers
Cellular Antennas, Mounts & Accessories, Ltd. 5299 West Washington Boulevard Los Angeles, CA 90016 213 936-0073 (tel) 800-634-3948 (toll-free) 213-934-0892 (fax)	Cellular Antennas Accessories
Cellular Communications Corporation 23161 Lake Center Drive, Suite 120 Lake Forest, CA 92630 714-859-3020 (tel) 714-830-4344 (fax)	Credit Card Phones Data Interfaces
Cellular Depot, Inc. 611 County Line Road, Suite D Huntingdon Valley, PA 19006 215-364-7879 (tel) 800-421-9175 (toll-free)	Wholesaler
Cellular Design Corporation 906 Long Island Avenue Deer Park, NY 11729 516-667-7447 (tel)	Cellular Accessories
Cellular International, Inc. 2030 Vineyard Avenue Escondido, CA 92025 619-740-0700 (tel)	Vehicular Antennas Cellular Accessories
Cellular Marketing Box 430016 Houston, TX 77243 713-468-1100 (tel)	Cellular Accessories
Cellular Telephone Sales of Colorado 102 East Cleveland Street Lafayette, CO 80026 303-758-1990 (tel)	Cellular Accessories

Manufacturer

Equipment or Service

Cellular Wholesalers
7855 Grosse Point Road, Suite N1
Skokie, IL 60077
312-982-6800 (tel)
800-542-2770 (toll-free)

Wholesaler
Cellular Accessories

Cellwave R.F.
Route 79
Marlboro, NJ 07746
201-462-1880 (tel)
201-462-6919 (fax)

Portable Antennas
Vehicular Antennas

Centurian International, Inc.
Box 82846
4555 North 48th Street
Lincoln, NE 68501
402-467-4491 (tel)
800-228-4563 (toll-free)

Portable Antennas

Childs Corporation
Route 1
Box 155
Janesville, MN 56048
507-267-4388 (tel)
800-533-0998 (toll-free)

Portable Antennas
Vehicular Antennas

Cincinnati Bell Information Systems
Cellular Business Systems Division
1661 Freehanville
Mount Prospect, IL 60056
312-299-9500 (tel)
800-327-3900 (toll-free)

Billing Systems

COM-SER Laboratories
Box 1776
Bradenton, FL 33506
813-355-2779 (tel)

Test Equipment

Communications Associates
Box 2399
305 Republic Avenue
Joliet, IL 60434
800-435-9313 (toll-free)
815-741-2152 (fax)

Wholesaler
Cellular Accessories

Manufacturer	Equipment or Service
Communications Insurance Consultants Box 2525 Shawnee Mission, KS 66201 913-362-3818 (tel)	Insurance
Communications Software, Inc. 2388 Pleasantdale Road Atlanta, GA 30340 404-448-5259 (tel)	Billing Systems
Comm 88 3750 Texas Avenue South Minneapolis, MN 55426 612-720-2469 (tel)	Cellular Accessories
CT Systems Box 470 5245 Hornet Avenue Beech Grove, IN 46107 317-787-5271 (tel) 800-245-6356 (toll-free) 317-788-4197 (fax)	Test Equipment
Curtis Electro Devices Box 4090 Mountain View, CA 94040 415-964-3846 (tel) 415-964-3574 (fax)	NAM Programmers
Cushman Electronics, Inc. 1525 Attenbury Lane San Jose, CA 95131 408-432-8100 (tel) 800-538-7020 (toll-free)	Test Equipment NAM Programmers
Custom Interface Accessories 2499 Lincoln Boulevard Marina del Rey, CA 90291 213-623-4242 (tel)	Cellular Accessories
DATA I/O 10525 Willows Road North East Redmond, WA 98073-9746 206-881-6444 (tel)	NAM Programmers

Manufacturer

Dataradio, Inc.
1819 Dorchester Boulevard West
Montreal, PQ
Canada H3H 2P5
514-932-6600 (tel)
514-932-7638 (fax)

DEBEfone, Inc.
6 Commerce Boulevard
Palm Coast, FL 32037
904-445-0304 (tel)

Decibel Products, Inc.
Box 47128
3184 Quebec Street
Dallas, TX 75247
214-631-0310 (tel)
214-631-4706 (fax)

DiamondTel Mitsubishi Electric Sales
800 Bierman Court
Mt. Prospect, IL 60056
312-298-9223 (tel)
800-323-4216 (toll-free)
312-298-0567 (fax)

Direct Enterprises
Box 311
Hartford, CT 06141-0311
203-549-5822 (tel)

Dova Controls
4052 148th Avenue
Redmond, WA 98052
206-881-7414 (tel)
206-883-9054 (fax)

Drive Phone
Box 0588
Paramus, NJ 07653-0588
201-843-6400 (tel)

Equipment or Service

Data Interfaces

Cellular Accessories

Base Station Antennas
Vehicular Antennas

Cellular Terminals

Cellular Accessories

NAM Programmers

Wholesaler
Cellular Accessories

Manufacturer | Equipment or Service

John Fluke Manufacturing
Box C9090
Everett, WA 98206
206-356-5295 (tel)

Test Equipment

Fujitsu America, Inc.
15707 Rockfield Boulevard
Irvine, CA 92718
714-837-0671 (tel)
714-837-0677 (fax)

Cellular Terminals

General Electric Company
Mobile Communications Business Division
Lynchburg, VA 24502
804-825-7000 (tel)

Cellular Terminals
Cell-Site Equipment

Glenayre Electronics
849 Industry Drive
Seattle, WA 98188
604-293-1611 (tel)

Cellular Terminals

Harada Industry
1650 West Artesia Boulevard
Gardena, CA 90248
213-532-1111 (tel)

Portable Antennas
Vehicular Antennas

Hewlett-Packard Company
1620 Signal Drive
Spokane, WA 99220
509-922-4001 (tel)

Test Equipment

Hirschmann of America, Inc.
Box 229
Riverdale, NJ 07457
201-835-5002 (tel)

Portable Antennas
Vehicular Antennas

Hitachi Sales Corporation
401 West Artesia Boulevard
Compton, CA 90220
213-537-8383 (tel)
213-515-6223 (fax)

Cellular Terminals

Manufacturer

Howe Industries
Box 1040
Sanford, FL 32772-1040
305-323-1830 (tel)

Hustler, Inc.
1 NewTronics Plaza
Mineral Wells, TX 76067
817-325-1386 (tel)
800-327-9076 (toll-free)

Intercontinental Development Corporation
25222 Richards Road
Spring, TX 77373
713-363-9466 (tel)

International Voice Products
Figgie International, Inc.
1849 West Sequoia Avenue
Orange, CA 92668
714-937-9010 (tel)
714-544-4045 (fax)

Kathrein Antennas, Inc.
26100 Brush Avenue, Suite 319
Euclid, OH 44132
216-289-1271 (tel)

Kraco
505 East Euclid Avenue
Compton, CA 90224
213-639-0666 (tel)
800-421-1910 (toll-free)

Larson Electronics
Box 1799
11611 North East 50th Avenue
Vancouver, WA 98668
206-573-2722 (tel)

Equipment or Service

Cellular Accessories

Vehicular Antennas

Cellular Accessories
Portable Antennas
Vehicular Antennas

Cellular Accessories

Base Station Antennas
Vehicular Antennas

Cellular Terminals

Vehicular Antennas
Mounting Hardware

Manufacturer

Lease Acceptance Corporation
7900 South Cass, Suite 160
Darien, IL 60559
312-960-9393 (tel)
312-557-9100 (fax)

Lone Star Antennas
Box 549
Prosper, TX 75078
214-347-2041 (tel)

Marconi Instruments
3 Pearl Court
Allendale, NJ 07401
201-934-9050 (tel)
201-934-9229 (fax)

Matt's Mobilephone
54 Tuxedo Avenue
Providence, RI 02909
401-273-4989 (tel)

Maxrad, Inc.
2495 Pan Am Boulevard
Elk Grove, IL 60007
312-595-3933 (tel)
800-323-9122 (toll-free)

Micro Office Technology, Inc.
Box 9955
College Station, TX 77840
409-696-1028 (fax)

Mitsubishi International
879 Supreme Drive
Bensonville, IL 60106
312-860-4200 (tel)

Mobile Executive Communications
8530 Wilshire Boulevard, Suite 404
Beverly Hills, CA 90211
213-275-7945 (tel)

Equipment or Service

Leasing Services

Base Station Antennas
Vehicular Antennas

Test Equipment

Cellular Accessories

Portable Antennas
Vehicular Antennas

Billing Systems

Portable Antennas
Cellular Terminals

Cellular Accessories

Manufacturer

Mobile Mark, Inc.
9001 Exchange Avenue
Franklin Park, IL 60131
312-456-0016 (tel)
800-648-2800 (toll-free)

Modublox & Company, Inc.
2167 Calle Guaymas
La Jolla, CA 92037
619-456-0016 (tel)
800-525-2283 (toll-free)

Morrison & Dempsey Communications
19201 Parthenia Street, Suite D
Northridge, CA 91324
818-993-0195 (tel)
818-993-7209 (fax)

Motorola, Inc.
1301 East Algonquin Road
Schaumburg, IL 60196
312-397-1000 (tel)
800-341-4430 (toll-free)
312-576-2702 (fax)

NBCC Cellular, Inc.
200 Park Central Boulevard, Suite 15
Pompano Beach, FL 33064

NEC America
1525 Walnut Hill Lane
Irving, TX 75038
214-580-9100 (tel)
214-550-8456 (fax)

Nokia-Mobira
2300 Tall Pines Drive, Suite 100
Largo, FL 33541
813-536-5553 (tel)
813-530-7245 (fax)

Equipment or Service

Vehicular Antennas

Portable Antennas
Vehicular Antennas

Data Interfaces
Cellular Accessories

Cellular Terminals
Cell-Site Equipment
Cellular Switching Equipment
Vehicular Antennas
Test Equipment
Mounting Hardware
NAM Programmers

Portable Antennas

Cellular Terminals
Test Equipment

Cellular Terminals

Manufacturer

Northern Telecom
1201 East Arapaho Road
Richardson, TX 75081
214-234-7500 (tel)

NovaTel Communications, Ltd.
1020 64th Street N.E.
Calgary, AB
Canada T2E 7V8
403-295-4500 (Can. tel)
817-332-3027 (U.S. tel)
403-295-0230 (fax)

OKI Telecom
Cellular Telephone Division
22-08 Route 208
Fairlawn, NJ 07410
201-654-1414 (tel)
201-734-1179 (fax)

ORA Electronics
Box 4029
20120 Plummer Street
Chatsworth, CA 91313
818-701-5848 (tel)
800-431-8124 (toll-free)
818-349-6887 (fax)

Panasonic Industrial Company
Two Panasonic Way
Secaucus, NJ 07094
201-348-7933 (tel)
201-348-7934 (fax)

PanaVise Products, Inc.
2850 East 29th Street
Long Beach, CA 90806
213-595-7621 (tel)

PCI
1021 12th Street, #204
Santa Monica, CA 90403
213-827-7892 (tel)
213-306-0119 (fax)

Equipment or Service

Cell-Site Equipment
Cellular Switching Equipment

Cellular Terminals
Cell-Site Equipment
Cellular Switching Equipment

Cellular Terminals

Portable Antennas
Vehicular Antennas
Mounting Hardware

Cellular Terminals

Cellular Accessories

Cellular Accessories

Manufacturer

Photocomm, Inc.
7735 East Redfield Road
Scottsdale, AZ 85260
602-948-8003 (tel)
800-223-9580 (toll-free)

Plexsys Corporation
Box 3217
Quincy, IL 62305
217-223-0692 (tel)

Power Tek Industries
14550 East Fremont Avenue
Englewood, CO 80112
303-680-9400 (tel)

Racom, Inc.
5504 State Road
Cleveland, OH 44134-2299
216-351-1755 (tel)

Radio Shack
Division of Tandy Corporation
1500 One Tandy Center
Fort Worth, TX 76102
817-390-3244 (tel)

Repco, Inc.
2421 North Orange Blossom Trail
Orlando, FL 32804
305-843-8484 (tel)
800-327-5633 (toll-free)
305-841-0331 (fax)

Ritron, Inc.
505 West Carmel Drive
Carmel, IN 46032
317-846-1201 (tel)
317-846-4928 (fax)

Equipment or Service

Solar Cells

Cell-Site Equipment
Cellular Switching Equipment

Portable Antennas
Cellular Accessories

Cellular Accessories

Cellular Terminals

Wholesaler
Cellular Terminals
Cellular Accessories

Data Interfaces

Manufacturer	Equipment or Service
Rohde & Schwarz Polrad, Inc. 5 Delaware Drive Lake Success, NY 11042-1116 516-328-1100 (tel) 516-352-1183 (fax)	Test Equipment
Rona Leather Box 372 1 Airport Industrial Campus Little Ferry, NJ 07643 201-807-0100 (tel) 201-807-0195 (fax)	Cellular Accessories
Schlumberger 22 Morgan Street Irvine, CA 92718 714-583-7735 (tel) 714-583-9086 (fax)	Test Equipment
Scientific Dimensions, Inc. Box 26778 2655 Pan American Freeway N.E. Albuquerque, NM 87125 505-345-8674 (tel) 800-523-6180 (toll-free)	Mounting Hardware
H. Scott Enterprises 500 Seventh Avenue New York, NY 10018 212-869-3838 (tel)	Cellular Accessories
Shure Brothers, Inc. 222 Hartrey Avenue Evanston, IL 60202-3696 312-866-2517 (tel)	Custom Hands-Free Equipment
Signal Measurement Company 8410 Prine Street Tomball, TX 77375 713-351-0700 (tel)	Mounting Hardware

Manufacturer

Trilectric, Inc.
Division of Celltronics, Inc.
10040 Mesa Rim Road
San Diego, CA 92121
619-587-0656 (tel)
800-551-8551 (toll-free)
619-587-0049 (fax)

Uniden Corporation of America
6345 Castleway Court
Indianapolis, IN 46250
317-842-0280 (tel)

Valor Electronics
185 West Hamilton Street
West Milton, OH 45383
513-698-4194 (tel)
800-543-2197 (toll-free)

Vision Systems, Inc.
1890 Maine Street
Quincy, IL 62301
217-228-0361 (tel)
217-228-9350 (fax)

Western Mobile Telephone
1924 South Anaheim Boulevard
Anaheim, CA 92805
714-774-0520 (tel)

White Automotive
17100 Southfield
Allen Park, MI 48101
313-388-8800 (tel)

Wiltron Company
490 Jarvis Road
Morgan Hill, CA 95037
408-778-2000 (tel)
408-778-0239 (fax)

Equipment or Service

Mounting Hardware

Cellular Terminals

Portable Antennas
Vehicular Antennas

Billing Systems

Hands-Free Equipment

Cellular Accessories

Test Equipment

Manufacturer

Wing Communications, Inc.
Box 4118
Hialeah, FL 33014
305-875-0770 (tel)

ZK Celltest Systems
108 East Fremont Avenue, Suite 80
Sunnyvale, CA 94087-3201
408-248-8832 (tel)

Equipment or Service

Cellular Accessories

Test Equipment

Index

Page numbers enclosed in brackets indicate illustrations.

A/B switch, 68, 132, 179, 201
access, 9, [10], 201
accessories, 74–77
   marine, 150–51
ACCOLC, 36
adapter, cigarette lighter, 62
adhesives, 125, 134
administration, 20, 22
Advanced Mobile Phone Service. *See* AMPS
airbag, 115–16
alarm, 24, 35, 101–2, 141, 161
alert
   horn, 37
   silent, 73–74
allocation, dynamic channel, 20
alphanumeric operation, 69
amperes, 201
amplifier, linear, 25, 208
amplitude, 201
AMPS, 3, 30–31, 201, 202, 204
   specifications, 31–32
analyzer, spectrum, 29–30
answering machines, 37
antennas, 15, 25–26, 34, 43–44, 64, 85–86, 162
   0–db, 182
   3–db, 182
   dropped calls, with, 181–86

elevated–feed, 88, [91, 92], 93, 109, [111], 144–45, 155, 181
elevated-feed, installing, 128–29
five-eighths-wave collinear, 91–92
glass-mount, [88, 89, 90], 91, 107, [108–9], 111, 181–82
glass-mount, installing, 124–28
installing, 124–30
location selection of, 107–13, 142
location selection of, marine, 145, [146–47]
Marconi-type, 85–86, [87], 91, 93
marine, 144–48
marine, installing, 147–48
quarter-wave, 85, 92, 182
radiation patterns of, 86, [87]
resonant, 85
roof-top, 62, [93–95], 96, [97–98], [112], 113, 129–30, 181
rural-use, 154–55, [156–57], 158
selection of, 88–98
space-diversity receive, 26, 207
stolen, 181
theoretical, [87]
theory of, 79–81
yagi, 155, [157], 208
Armor All™, 137, 140
assignment, channel, 23
   manual, 32

AT&T, 208
audio, 56
　quality, 2, 3
　testing, 29
　voice, processing and conditioning
　　of, 24
autotest, 30–32, 196

batteries, 56, 60
　backup, 201
　disconnecting, 116, 142
　gel cell, 61
　internal, 61–62
　lead-acid, 61, 154, 204
　lithium, 115, 190, 198
　marine deep discharge, 159
　nickel-cadmium, 61, 205
　safety with, 115, 116, 168
　solar, 158–59
　storage, 167
baud, 201
Bell Companies, 205, 208
Bell Laboratories, 2
Bell Telephone, 2
billing, 3, 18
binary, 35
bits, 35–37, 201–2
　busy/idle, 14–15, 202
　repertory memory. See REP
blank and burst, 202
blocking, 9
BMW, 111, 114
boards
　audio, 170–71
　circuit, 180
　control head, 170–71, 197
　logic, 170–71, 190, 197
　power, 170
　receiver, 170–71
　synthesizer, 170–71
　transmitter, 170–71, 197–98
Bridge™, 162
button
　clear, 73
　recall, 67
　SEND, 68, 78, 171, 175
bytes, 35, 202

cable
　antenna, [183]
　coaxial, 83–84, 86, 148, 164–65
　copper, 204
　data, 55, 57, 140, 142
　fiber optical, 20, 22–23, 204
　heliax, 26
　power, 56–57, 116, [117], 118–20,
　　142, 166–67
　RG 58U, 83
　running, 118–20, 125–26, 129,
　　139–40, 164, [165]
　"T", 71
call processing, 9–11, 22, 33
calls
　dropped, 11–12, 15, 180–89
　test, 140, 142
Canada, 35, 204, 205
carrier
　alternate, 131, 172, 173, 196
　FM, 25
　nonwireline, 37, 68, 205
　service, 34, 36, 77–78
　wireline, 37, 68, 208
carrier-to-interference ratio, 16, 202
carrier-to-noise ratio, 16
carry bag, 63
cathode ray tube. See CRT
cell site, 5, 9, 11, 12, 16, 20, 22, [24],
　　25, 155
　coverage, [12]
　power output, 6
　system description, 23
Cellular Geographical Service Area.
　　See CGSA
Cellular One, 68, 181
cellular system, typical, 5, [6]
cellular telephones
　features of, 67–74
　grounding, [119]
　hand-held portable, 58, [59],
　　63–67, 144
　mobile-only, 55–59, [56, 57]
　rural, 154–55
　subsystems, 170–71
　types of, 5, [7]
cellular terminal, 5
cellular test center, 28, [29], 30-34,
　　[172]
CGSA, [8], 9, 11, 23, 202
channel seizure, 14–15
channels, 2–3, 9, 20
　access, 9, 25, 201
　control, 23, 202, 204

INDEX

expanded, 68
forward voice, 204
initial paging, 36
paging, 9, 25, 206
reverse control, 206
reverse voice, 207
voice, 9, 208
charger
  AC, 62
  battery, 22, 45,
chassis, control, 22
checklist
  installation, 51–52, 100
  preinstallation, 100–1
circuits
  logic, 190
  open, 84
  short, 83
  timing, 22
circuitry, channel-grabbing, 2
class
  access overload, 36
  mobile station, 205
cleaner, flourocarbon, 198
clips, cable, 125
CMOS, 176, 190
code
  digital color, 203
  Manchester, 18, 205
coding
  BCH (Bose-Chaudhuri-
    Hocquenghem), 18, 201
  computer logic, 3
  international, 35
collinear radiator, 88
collision, channel, 202
combiners, RF, 26
companding, 24–25, 202
  syllabic amplitude, 207
complaints, customer, xi
computers, 49–53, 99
  on-board, 115
conferencing, 20
connections
  ground, [179]
  T1, 19
connectors
  3M, 121, [122], 208
  antenna, 197
  coaxial, 126, [127], 165

crimp-on, 46
solderless, 46, 118
spade, 118, 123
constant power wire, 122–23
consumption, current, 60
control
  data and voice-link, 24
  dynamic power-level, 34, 203
  power level, 23
  transmission, 23
  volume, 73
control channels, 9, 11
control head, 55–57, 64, 69–71, [137], 138, 197
  console-mount, [102]
  dashboard-mount, [103], 138
  floor-mount, [103]
  hands-free, 140
  location of, 102–3
  marine, 144
  mounting, 135–38, 142, 148, [149]
convertibles, antennas with, 108, 113, 139
Corvette, 113, 145
counter, frequency, 28
coupler design, high-impedance, 90
coupling box, installation of, [110], 124–25
crimpers, 47, 126
crimping, 84, 126–27
CRT, 29
current, DC, [185]
current/phase relationship, 202
customer, checking out, 141, 142

damage
  salt, 106
  water, 191–92, 198
data base, system, 22
data bus, 160, 202
data networks, 5
DBX™, 25, 202
"dead spots," 15
defoggers, rear-window, 108–11
demodulate, 16
design, 23
device, voting, 26
diagnostics, 20
dielectric, 203
digit, signaling, 19

directed retry, 12–14, 203
directory, negative, 174
disconnect, 203
display, 56
  backlit, 68–69
display number
  LCD, 72–73
  LED, 72–73
  own, 74
distribution
  rural power, [167]
  rural signal, [167]
diversity
  frequency, 26
  time, 26
Dolby™, 25, 203
drills, 44, 46
driver set, hex, 43
DTMF, 32, 37, 74, 203
dual power, 33
dual registration, 35
Dual-Tone Multi-Frequency. *See*
  DTMF

echoes, acoustic, 15, 17
EE, 37
EEPROM, 38, 203
effect
  end, 86
    Losee-Shosteck, 96
    skin, [184]
effective radiated power. *See* ERP
electrical check, preinstallation,
  99–102, 141
electronic serial number. *See* ESN
electrons, 203
END key, 203
end-to-end signaling. *See* EE
epoxy, 194–95
equalization, 17
equipment
  cellular telephone, 55–78
  computer, 49–53
  control-channel, 25
  locating-channel, 25
  power-distribution, 22, 25
  test, 27

trunk-interface, 25
voice-channel, 25
erlang, 203
ERP, 6, 203
errors, data, 98
ESN, 18, 20, 173–74, 196
eye protection, 44

facilities, installation, 47, [48], 49
"fax" machine, 74, 151, 162
FCC, 2, 3, 5, 68, 153
Federal Communications
  Commission. *See* FCC
fees, determining marine installation, 143–44, 147
fender covers, 44
flag, 203
  local-use, 35
format, Manchester, 18, 205
formatting, paging channel message,
  23
forwarding, call, 20, 77
frequency, 204
frequency counter, 28
frequency shift keying. *See* FSK
FSK, 18, 204
functions, telephone switching, 22
fuses, 176, [177], 179, 197

gain, 204
gas tank, leak in, 193, [194]
generator, signal, 28
GIM, 37
goggles, 44
Great Britain, 207
grinder, 46
ground plane, 85, 91, 93
ground wire, 116, [117–18]
gun, heat, 47

handoff, 11–12, [13], 16, 20, 23, 32,
  98, 204
handset, 60, 69
  programming, 39–40
hands-free, 37, 69, 138–40, 192–93
hardware, 2, 22, 45
Hercules™, 51

hex driver set, 43
hole punch, 45, 129, 134
  kit, 43
horn alert, 37
"hot wire," 150

IBM™, 51
identification
  group, 204
  transmission, 32
ignition sense wire, 120–21, [122], 177
impedance, 204
Improved Mobile Telephone Service. *See* IMTS
IMTS, 1, 2, 204
indicator
  call in-absence, 73
  signal-strength, 71, 155, 207
installation procedures, review of, 141–42
installations, 27–53
  beginning, 114–16
  checking, 128
  fixed, 153–67
  marine, 143–51
  mobile, 86
  on-site, 47
  planning vehicle, 102–13
  repairing, 169–99
  rural, 153–67
  troubleshooting, 169–99
  vehicle, 99–142
integration, large-scale. *See* LSI
Intel, 51
interface
  AC, 22
  computer, 154, 162
  data-link, 25
  landline, 19
  man/machine, 22
  RJ11, 74, [75], 150, 154, 160, [161], 162
interference
  co-channel, 15
  multipath, 16, [17], 95
IPCH, 36

Jaguar, 114, 121

keypad, 56, 130
  backlit, 68–69
keys, transceiver, 47
kick panel, [119]
kit
  amplifying, [65, 66], 67
  limo, [70]
  mobile installation, 62–63
  nonamplifying, 63, [64]
  vehicle installation, [67]

landlines, 20–22, 204, 206
light
  shop, 45
  test, [40], 41, 84, 120, 176
limousine option, 70
line
  fuel, [195]
  power, [124], 176, [178]
  transmission, 81–84
links, data, 20
list, neighbor, 13
loading, system, 16
lock, electronic, 71–72, 189–90, 198, 203
Locktite™, 148
logic, computer, 3
Losee–Shosteck effect, 96–98
LSI, 60, 204
Lynx, 174

machines, answering, 37
mail, voice, 5, 37, 78
maintenance, 20, 22
manager, service, 99, 102
mark
  group identification, 37
  station class, 35–36
matrix, switching, 22
MCI, 74
measurement, power, [185]
megahertz, 205
memory
  electronically erasable programmable read-only. *See* EEPROM

memory (*continued*)
  programmable read-only. *See* PROM
  read-only. *See* ROM
  repertory, 122
  speed dial, 67
Mercedes-Benz, 43, 46, 109, 111, 114–15
message formatting, paging-channel, 23
messaging systems, 5
meter
  power, 28
  watt, 34, 184, [186]
microphone, hands-free, 56, 69, [138–39], 140, 142, 199
  calibrating, 140
  location of, 138–40
microprocessor, 205
microwave stations, 20, 22–23
MIN, 35, 205
mispages, 18
mobile identification number. *See* MIN
mobile phones/terminals, 5, 9, 11
Mobile Telephone Service. *See* MTS
Mobile Telephone Service Office. *See* MTSO
modem, 51
  cellular, 75, [76], 162, [163–64]
modulate, 16, 18
modulation, 82
  Pulse Code. *See* PCM
  digital, 203
modulator, 29, 205
modules, interface, 22
MSA, 205
MS-DOS™, 51–52
MTS, 1, 205
MTSO, 5–6, 11, 12–16, 18, [19], 20, [21], 22, 23, 25, 52, 173–74, 188, 205, 203, 206
multimeter, 41, 183–84
multipath, 17
multiplex, 16
Multiplexing, Time Division. *See* TDM
mute switch, 71

NAM, 33–40, [38], 130, 189, 193, 198, 203, 205–6
  programmer, 38, [39]
  programming, 34–37
NAM capability, dual, 35, 36
network
  data, 5
  switching, 19
NMT, 205
no-answer transfer, 73, 78
noise, 170, 205
  intersyllabic, 204
nonwireline, 37, 68, 205
Nordic Mobile Telephone system. *See* NMT
NPA, 9
number assignment module. *See* NAM
number display
  LCD, 72–73
  LED, 72–73
  own, 74
Number Planning Areas. *See* NPA
NYNEX, 68, 174

ohm, 205
Ohm's law, 84
oscillator, internal, 82
oscilloscope, 29–30
outpulse, 203, 206
overload, paging channel, 14
overview, system, 5

paging, 9, [10], 14, 206
parameters, 35, 36
patterns, radiation, 86, [87]
PCM, 20, 206
photovoltaic cells, 154, 158–59
Plain Old Telephone Service. *See* POTS
plane, ground, 85, 91, 93
plate, transceiver mounting, [135]
plates, coupling, 88–91
pliers, 45
plumber's snake, 43, 119, 121, 123, 130
PO2 service, 9
polarity, 206

INDEX 235

Porsche, 53, 111
port, RS-232, 162
POTS, 5, 14, 19, 35, 74, 160, 203, 206
power
 AC, 201
 DC, 202
 distribution, DC, 22
 forward, 186
 irregular, 190–91
 meter, 28
 output, 6, 61, 63
 source, 58, 61–62
 supply, 40–41, 62
powerboats, 145, 148
preinstallation electrical check, 99–102, 141
printer, 51
processing, call, 9–11, 22, 33
processor, 22, 56, 206
program, test center, 28
programming, NAM, 34–37
programs, computer, 5, 22, 51–53
PROM, 38, 130, 189, 206
propagate, 79, 206
protection, eye, 44
PS, 37
PTSN, 5–6, 19–20, 22, 206
Public Telephone Switching Network. *See* PTSN

quartz crystal, 82

radiator, collinear, 88
radio, 56
 security, 114–15
 transceivers, 9
 two-way, 1
radio band
 150-MHz, 1, 2
 450-MHz, 2
 800-MHz, 9
radio common carrier. *See* RCC
radio frequency. *See* RF
RAM, 206
range, 16
ratio
 carrier-to-interference, 16, 202
 carrier-to-noise, 16
 signal-to-noise, 12, 25
RCC, 2
RCL button, 67
receiver, 206
recognition, voice, 76–77
record, installation work, 99–100
records, keeping, 50, [52]
redundancy, 22, 206
registration, 206
 dual, 35
REP, 37
repertory memory bit. *See* REP
request
 flash, 204
 release, 206
resistance, 206
restraint system, passive, 115–16
retry, directed, 12–14, 203
RF, 11, 15, 22–23, 25, 40, 83, 88, 90–91, 95, 123, 188, 207
 combiners, 26
RJ11 interface, 74, [75], 150, 154, 160, [161], 162
RMS, 207
 deviation, 25
ROAM light, 68
roaming, 18, 207
Rolls-Royce, 111, 128, 182
ROM, 18, 207
RS-232 port, 162
RSA, 153
RSSI, 32, 187, 207
Rural Service Area. *See* RSA

Saab, 114
SAE, 43, 207
sailboats, 145, 147–48
salt, damage from, 106
SAT, 12, 23, 31–32, 97, 207
SCM, 36
screwdrivers, 42, 44
seizure, channel, 14–15
SEND button, 68, 78, 171, 175
service, PO2, 9
services, carrier, 77–78
short circuit, 83–84
Shosteck, Herschel, 96, 180

shutoff, auto, 63
SIDH, 35–37, 207
signal
   amplitude modulated, [30], [31]
   determining level of, 155
   generator, 28
signaling, 3, 17, 22, 25
   end-to-end. *See* EE
signal-to-noise ratio, 12, 25
silicone, 44, 118, 164
socket set, 43
software, 19, 22
solar panels, 154, [158], 159, [160], 165, [166], 167
Span™, 162
speaker
   call-monitoring, 69
   external, 69
   hands-free, 56, 69
specifications, AMPS, 31–32
spectrum, 3, 68, 207
   analyzer, 29–30
Spectrum Cellular, 162
spike, power, 2
spot remover, 46
Sprint, 74
stock quotes, 20
stream
   data, 202
   paging/access, 14, 206
subscriber, 34
subscriber verification, dynamic, 18, 174
subsystems, cellular telephone, 170–71
supervision, call, 202
switch, 207
   A/B, 68, 132, 179, 201
   mute, 71
   power, 176, 197
   system select, 68
switching matrix, 22
system
   antenna, 81–82, [83]
   cellular, 202
   operators, 2
   overview, 5
   preferred. *See* PS
System, Total Access Communications. *See* TACS

T1 connections, 19, 207
TACS, 207
tar, roof, 164, 165, 166
TDM, 20, 207
telephone service, integrated, 1
telephones, cellular, 55–78
   hand-held portable, 58, [59], 63–67, 144
   mobile-only, 55–59, [56, 57]
   transportable, [58], 58–63
temperature, operating, 73
terminal
   cellular, 5
   fixed, 5
   home mobile, 204
   mobile, 5, 14–16, 23–24
   portable, 5
test
   auto, 131
   manual, 32
   manual handoff, 130
test center, cellular, 28 [29], 30–34, [172]
testing, 27–47
time, operating, 60
timer, call, 69–70
tone
   ringing, 71
   test, 140–41
tools, 42–47
torx bits, 47
Touch-Tone™, 74, 203
traffic, 20, 22
transceiver, 55–57, 70–71, 142, 186–87, 189, 197, 208
   driver's-seat location, 106, [107], 136
   hatchback location, 107
   keys, 47
   location selection, 103–7
   mounting, 103–4, 132–35, 149–50
   trunk location, [105–6]
transfer, no-answer, 73, 78

transmitters, 29, 64, 81, 166, 208
    high-power, 1
    low-frequency, 2
    low-power, 5
trunks, voice, 19

U.S., 35, 201, 204–5
utility closet equipment, 166–67

vacuum, shop, 45
velcro, 134–35, 137–38, 139, 150
verification, dynamic, 18, 174
verification, electronic serial number.
    See ESN
video, 151
vise, 46
voice
    channels, 9, 12, 33
    mail, 5, 37, 78
    recognition, 76–77
    trunks, 19
voltage, 208
    regulator, 154, [159], 166–67, 191, 199

volume adjustments, 56
Volvo, 114

waiting, call, 20, 78
waiting area, [50]
warranty, 106, 192
water damage, 191–92, 198
watt, 208
watt meter, 34, 184, [186]
wave, sine, 82
wavelength, [81]
waves, 79, [80]
WD-40™, 182
weight, 60
wheel ramps, 45
whip, antenna, 182
window tinting, aftermarket, 109–10
wire
    constant power, 122–23
    ground, 116, [117–18]
    ignition sense, 120–21, [122], 177
wire cutters, 46
wireline, 37, 68, 208